1st International Conference on Recent Advancements in Design and Manufacturing (ICRADM 2020)

Held online due to COVID-19

IOP Conference Series: Materials Science and Engineering
Volume 1004

Surat, India
16 - 17 July 2020

ISBN: 978-1-7138-3316-1
ISSN: 1757-8981

Printed from e-media with permission by:

Curran Associates, Inc.
57 Morehouse Lane
Red Hook, NY 12571

Some format issues inherent in the e-media version may also appear in this print version.

No changes have been made to the content of these proceedings. There may be changes to pagination and minor adjustments for aesthetics.

Printed with permission by Curran Associates, Inc. (2021)

Institute of Physics
Dirac House, Temple Back
Bristol BS1 6BE UK

Phone: 44 1 17 929 7481
Fax: 44 1 17 920 0979

techtracking@iop.org

Additional copies of this publication are available from:

Curran Associates, Inc.
57 Morehouse Lane
Red Hook, NY 12571 USA
Phone: 845-758-0400
Fax: 845-758-2633
Email: curran@proceedings.com
Web: www.proceedings.com

TABLE OF CONTENTS

IOP Conf. Series: Materials Science and Engineering 1004 (2020) 011001 doi:10.1088/1757-899X/1004/1/011001

1st International Conference on Recent Advancements in Design and Manufacturing (ICRADM- 2020) from 16-17 July 2020

Preface

We are very delighted to introduce the proceedings of our **1st International Conference on Recent Advancements in Design and Manufacturing (ICRADM-2020)** held in virtual mode at textile and diamond city, Surat inside beautiful campus of Sardar Vallabhbhai National Institute of Technology (SVNIT). The institute was initially established as Sardar Vallabhbhai Regional College of Engineering & Technology in 1961 and elevated as a National Institute of Technology in October, 2002. It has been accorded the status of an Institute of National Importance vide NIT Act 2007. SVNIT is one of the pioneering engineering institutions of the country which has nurtured many outstanding engineers in India & abroad.

Initially ICRADM-2020 organized to exchange the innovation in field of thermal, design, manufacturing, materials into the scientific community and researchers along with the encourage the student to work in these field for the societal and research benefits. At the first proposed schedule of ICRADM-2020 outbreak of COVID-19 takes place and eventually, conference get postponed with new schedule. Though the virtual conference was new to us as well as our participants, we all made it successful, by proper management and enthusiastic participation of every person who involved directly or indirectly into it.

The inaugural function of the conference was enlightened by the august presence of **Prof. Harish Hirani**-Director, central mechanical engineering research institute (CMERI, Durgapur), **Dr. Kuldeep Mistry**- product development specialist, TIMKEN, Ohio (USA), **Dr. S. R. Gandhi**-Director, SVNIT, Surat (GJ), HOD Mechanical, faculty

IOP Conf. Series: Materials Science and Engineering 1004 (2020) 011001 doi:10.1088/1757-899X/1004/1/011001

1st International Conference on Recent Advancements in Design and Manufacturing (ICRADM- 2020) from 16-17 July 2020

members and more than 60 participants. Experts from different part of the country in the field of Mechanical, Material Science, Tribology and Manufacturing attended the ICRADM-2020 as well as given a very valuable speech during the virtual ICRADM-2020. **Prof. Harish Hirani** addressed on design thinking to rural sustainable development. His speech followed by the **Dr. Kuldeep Mistry,** guided the audience on topic, recent advancement in industrial tribology. Second day of conference started with the key note address of **Dr. Anand Dhruv**- Vishwakrma Government Engineering College, Chandkheda (Gandhinagar) on topic recent trends in micro forming. Thereafter **Dr. Dattaji Shinde** - VJTI, Mumbai elaborated the effect of electrospun nanofibers and carbon nanotubes on the properties of polymer composite and its failure analysis. The last but the most awaited address was delivered by **Dr. Umang Soni** - Netaji Subhas University of Technology, New Delhi on topic intelligent manufacturing: prospects of artificial intelligence in pandemic. The address was truly captivating and thought provoking. Peer reviewed accepted papers in conference presented by participants given a new idea, direction and inspiration to other listeners. So as a coordinator(s), we can say that, ICRADM-2020 provided an international forum to present, discuss, and exchange new, innovative ideas in recent advancements in design and manufacturing.

In the event, the conference was highly successful. Total 50 papers get presented in conference, which we feel good number during this epidemic COVID corona situation. After peer review process, papers get shortlisted and will be published with IOP-MSE (Scopus Indexed). Presented papers maintained the high potential recommended by the written abstracts and the programme was chaired in a

1st International Conference on Recent Advancements in Design and Manufacturing (ICRADM- 2020) from 16-17 July 2020

professional and efficient way by the session chairmen who were selected for their international standing in the subject.

Finally, it is appropriate that we record our thanks to our fellow members of the technical organizingcommitteefor their work in securing asubstantial input of papers and smooth conduction of (virtual) ICRADM-2020.We are also grateful to those who served as chairmen,without their support, the conferencecould not have been the success that it was. We also acknowledge the authors themselves, withoutwhose expert input there would have been no conference. We also acknowledge the important contributions of Director (SVNIT), Chairmen (TEQIP) and our Mechanical engineering department for providing such a huge platform and support.

Dr. Sandeep Soni, Dr. B. M. Sutaria and Dr. Achchhe Lal
Organizing Secretary (ICRADM - 2020)
Department of Mechanical Engineering,
Sardar Vallabhbhai National Institute of Technology (SVNIT),
(An Institute of National Importance, Government of India)
Ichchhanath, Surat -395007, (Guj.), India.

IOP Conf. Series: Materials Science and Engineering 1004 (2020) 011002 doi:10.1088/1757-899X/1004/1/011002

Peer review declaration

All papers published in this volume of **IOP Conference Series: *Materials Science and Engineering*** have been peer reviewed through processes administered by the Editors. Reviews were conducted by expert referees to the professional and scientific standards expected of a proceedings journal published by IOP Publishing.

- **Type of peer review: Single-blind**
- **Conference submission management system: Virtual/Online**
- **Number of submissions received: 49**
- **Number of submissions sent for review: 49**
- **Number of submissions accepted: 25**
- **Acceptance Rate (Number of Submissions Accepted / Number of Submissions Received X 100): 51**
- **Average number of reviews per paper: 2**
- **Total number of reviewers involved: 10**
- **Any additional info on review process: NIL**
- **Contact person for queries:**

Dr. Sandeep Soni,
Assistant Professor,
Department of Mechanical Engineering,
Sardar Vallabhbhai National Institute of Technology (SVNIT),
(An Institute of National Importance, Government of India)
Ichchhanath, Surat -395007, (Guj.), India.
Mobile No. +91-9727337702; LL- 0261-2201848
E-mail: s.soni@med.svnit.ac.in; ssonimed2009@gmail.com

IOP Conf. Series: Materials Science and Engineering 1004 (2020) 012001 doi:10.1088/1757-899X/1004/1/012001

To Study effect of VARTM process parameters for composite strength: Taguchi approach

A T Bhatt[1]* and P P Gohil[1]

[1]Faculty of Technology and Engineering, The M.S. University of Baroda, Vadodara, Gujarat, India.
* E-mail: alpa1109@yahoo.com

Abstract: Vacuum assisted resin transfer moulding (VARTM) has become well known fibre reinforced composite (FRC) manufacturing technique for various applications like boat hull, automobile parts, wind blades, aircraft parts etc. As name suggests, vacuum is used to assist resin to flow inside fabric during impregnation. The flow of resin happens due to the pressure difference created due to vacuum supply pipe (at vacuum pressure) and resin supply pipe (at atmospheric pressure). The flow of resin bonds the fabric to make a solid part. Various parameters affect the quality of part produced by VARTM. In this paper effect of three process parameters were studied, which include number of layers, position of resin supply and vacuum supply location. Each parameter with three levels was studied to find the effect of these parameters on tensile strength and flow velocity. Jute and polyester resin were selected as material to prepare laminates. The experiments were performed as per Taguchi Orthogonal Array L(9). It has been observed that the mechanical properties were best achieved when position of resin supply was at middle, vacuum supply location was from edge to centre and numbers of layers were five.

1. Introduction

Vacuum Assisted Resin Transfer moulding a very old method, was known as vacuum infusion method. It was used to manufacture boat hull in 1950s [1] but over the period, due to limitations of this method, this method was called trial and error method to manufacture fibre reinforced composite parts. Details of VARTM process, its application and its variants were discussed by suumerscale and Searle [2] in his review. VARTM history, development and future trends have been discussed by Bhatt et al. [3]. As shown in figure 1, in VARTM resin was infused inside the layers of fabric due to pressure difference created between resin supply pipe (at atmosphere pressure) and vacuum pipe (at vacuum pressure set by vacuum pump). The fabric was laid on glass table, after applying three coats of release spray. The layers fabric were covered with Peel Ply, High Permeable Media (HPM) and vacuum bag and sealed by sealant tape. From one end vacuum was created and from other side resin was supplied.

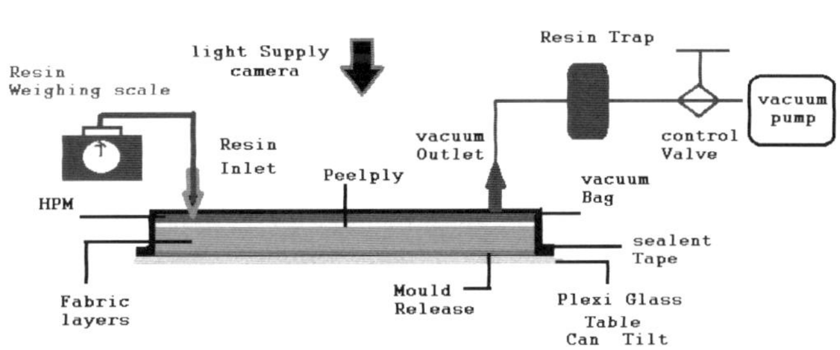

Figure 1: Schematic view of developed VARTM setup.

Various factors affect the quality of part produced by VARTM, Kuentzer et al. [4] wrote about how to control void content by bleeding the resin flow which ensures flow between and within tows. Kadari et al. [5] found that higher vacuum, high mould temperature and low inlet resin pressure can increase fibre volume fraction and reduce void. Rodriguez et al. [6-7] talked about unsaturated and saturated permeability of jute fabric. The permeability decreases during unsaturated condition due to fibre absorption and permeability also decrease during saturated flow due to fibre swelling. It was observed that jute absorbs more resin then glass. Saturated permeability was higher than unsaturated permeability. Tripathi et al. [8] have discussed permeability analytically, numerically and experimentally to measure the flow rate. Flow compaction behaviour was studied by Grimsley et al. [9] They studied dry and wet compaction of fabric. The fabric dry compaction took place while applying vacuum. When resin was introduced to fabric wet compaction occurred. Due to fabric lubrication, laminate thickness reduced. After which, resin pressure increased and there was spring back effect in laminate thickness. Laminate final thickness varied till resin curing took place. To improve the dimension tolerance of laminate during VARTM Gama et al. [10] have performed debulking cycle to get better compaction. However to get uniform thickness it was suggested to refer Closed-Micro-Flow where in uniform thickness with high flexural strength may be achieved by keeping vacuum line and infusion line closed after infusion and giving additional compaction by secondary vacuum line. Yoon et al. [11] performed experiments to find effect of gravitation, permeability, and resin supply pressure and tube head loss on filling time. The maximum filling time was achieved when the flow was against gravity. It was studied that by using longer pipe, filling time increases and applying pressure more than atmospheric pressure, chances of vacuum leakage increases. Rigas et al. [12] performed numbers of experiments to understand the effect of resin supply location, distribution media and cyclic load with double vacuum bag on fibre volume fraction. Heider and Gillespie [13] have commented that due to pre-infusion, de-gasification, fabric debulking and fabric drying void content will minimise. The HPM, resin pressure, fabric porosity will help to get 3D flow during impregnation. They also talked about dual state flow and effect of capillary to increase microscopic flow. Ghabezi et al. [14] have worked on effect of resin inlet tube diameter and suggested to use tube diameter 1.5 to 2 times the thickness of laminate.

2. Parameter involved in VARTM process

Various parameters affect the mechanical properties of laminate, thickness uniformity and fibre volume fraction. The most important were pressure, temperature, viscosity, permeability, volume fraction, filling time, angle of attack of the nozzle, orientation of fabric layers etc. Cause and effect diagram prepared to show various factors responsible for getting variation in part thickness, fibre volume fraction and tensile strength of FRP composite laminates is depicted in figure 2.

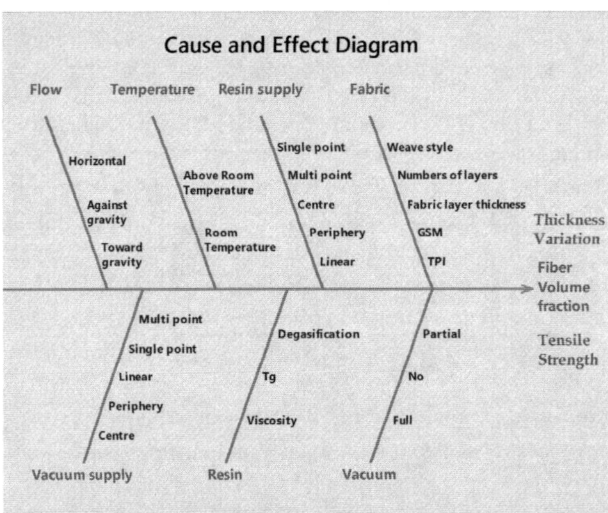

Figure 2: Cause and effect diagram for fiber volume fraction.

In this paper VARTM process has been studied using some of the parameters causing variations in output. This includes a) numbers of layers (4, 5, and 6), b) position of resin supply (Top, Bottom, and Middle) c) vacuum supply location (EC – edge to centre, CE- centre to edge, LR- left to right). Figure 3 shows the detail of all three arrangements.

Figure 3: Location of vacuum supply (EC, CE, LR).

3. Materials and Methods

From table 1, for 3 variable and each variable with 3 levels, Taguchi L(9), orthogonal array has been used to study the effect of variables on tensile strength, weight fraction and thickness variation. Taguchi Orthogonal Array L(9) method is used to study the effect of more than one variable for given output with less numbers of experiments then full factorial design. Taguchi method can be used to study which parameter causes more variation in output than the other and which level gives better result for a given parameter [15]. Table 2 shows numbers of experiments to be conducted for parameters shown in table 1, with Taguchi Orthogonal Array L(9) design. The experiments are performed not as per the standard order but as per run orders to remove biasness in the experiments.

Table 1: Parameter matrix.

Level	Parameters		
	Number of Layers	Position of Resin Supply	Vacuum supply location
1	4	T (Top)	EC (Edge to centre)
2	5	M (Middle)	CE (Centre to edge)
3	6	B (Bottom)	LR (Left to right)

Table 2: Design matrix – Taguchi L(9).

Run Order	Std. Order	Parameters		
		No. of Layers	Resin Position	Vacuum supply location
4	1	4	Top	Edge Centre
5	2	4	Bottom	Centre Edge
1	3	4	Middle	Left – Right
6	4	5	Top	Centre Edge
10	5	5	Bottom	Left – Right
7	6	5	Middle	Edge Centre
11	7	6	Top	Left – Right
8	8	6	Bottom	Edge Centre
9	9	6	Middle	Centre Edge

4. Developed experimental setup

The experimental setup was developed to perform experiments as shown in figure4. Plunkeet [16] has developed similar type of setup. Special arrangements have been made to ensure supply of resin 500 mm above and 500 mm below the level of table. C-Frame has been arranged on table to locate the camera and lamp. As shown in figure 4, resin is supplied from top (500 mm above the table level), vacuum from periphery and resin from centre. Table 2 shows standard order 1, Run order 4 to correlate the below arrangement.

Figure 4: Developed experimental setup.

The resin is taken 2 times the weight of fabric. Jute is taken as reinforcement as it is ecofriendly and less number of experiments has been performed with VARTM process and polyester with MEKP (1%) and cobalt (0.5%) as matrix to make laminate. The resin, hardener and accelerate ratio is so chosen as to ensure enough time for impregnation before geletion. Similar fibre matrix combinations have been used by other experts as well. [17-20]. Size of laminate was kept 300 mm X 300 mm. Before starting the impregnation following preliminary activities have been done. (i) Application of three coats of release spray (ii) weighing of fabric layers (iii) weighing of resin two times more than fabric layers (iv) readiness of vacuum bagging and vacuum leak check. During impregnation flow velocity of resin has been observed by camera. After impregnation, (i) vacuum pump was kept on for 3 hours (ii) room temperature curing was done for 24 hours for each laminate (iii) thickness was checked at 25 locations, for each laminate (iv) all laminate were weigh after curing (v) tensile test as per ASTM 3039 [21-23] was done for all laminates (vi) fibre weight fraction was obtained by solvent method.

5. Results and Discussion

5.1 Tensile strength and flow velocity

Table 3 and figure 5 show effect of various parameter and their levels on tensile test. As per this the tensile test was maximum when number of layers are 5, position of resin supply is at middle and vacuum supply location is from edge. ASTM D3039 was used to measure the tensile strength of laminate. 250 mm X 25 mm laminate were cut from each laminate to find the value of tensile strength. It is also important to notice that vacuum supply and number of layers are more important parameters than resin supply location to get the maximum tensile strength.

Table 3: Tensile strength and flow velocity.

Std. Order	No. of Layers	Resin Position	Vacuum supply location	Tensile stress, N/mm2	Flow velocity mm/s
1	4	Top	Edge Centre	63.12	6.25
2	4	Bottom	Centre Edge	48.63	5.55
3	4	Middle	Left – Right	47.61	8.33
4	5	Top	Centre Edge	52.56	4.32
5	5	Bottom	Left – Right	59.23	7.16
6	5	Middle	Edge Centre	63.41	3.33
7	6	Top	Left – Right	29.53	8.13

8	6	Bottom	Edge Centre	50.04	6.16
9	6	Middle	Centre Edge	54.20	5.83

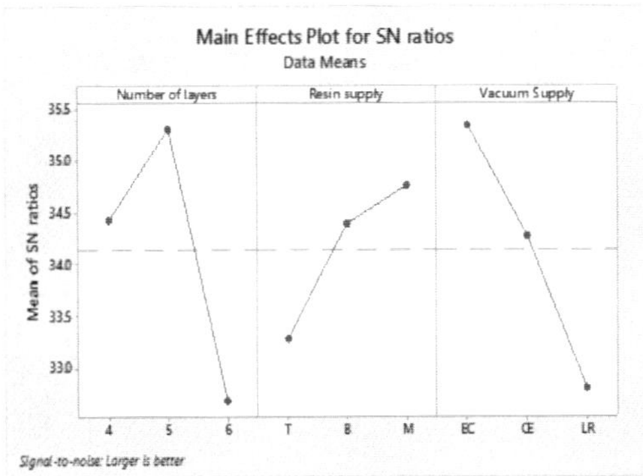

Figure 5: Effect on tensile strength (Max. is better)

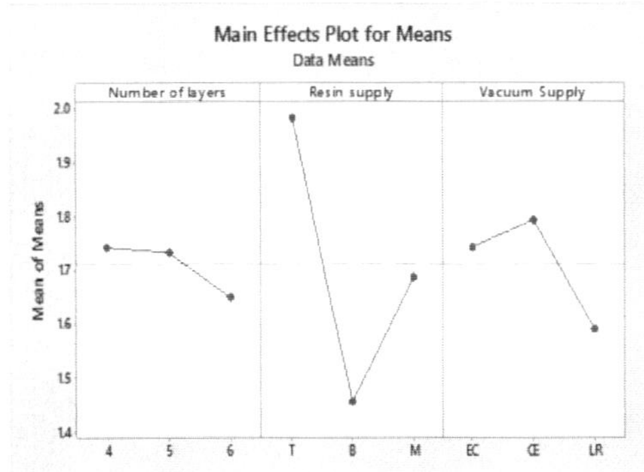

Figure 6: Effect of flow velocity (Max is better)

Table 3 and figure 6 show effect of various parameter and their levels on flow velocity. As per this figure flow velocity is maximum when number of layers is four, resin supply is from top and vacuum supply is from centre. The resin supply location is an important parameter to increase the flow velocity. This shows that flow velocity maximum does not ensure high strength in laminate. The quality of laminates depends on how well the resin is equality distributed between and within tows.

5.2 Variation in part thickness

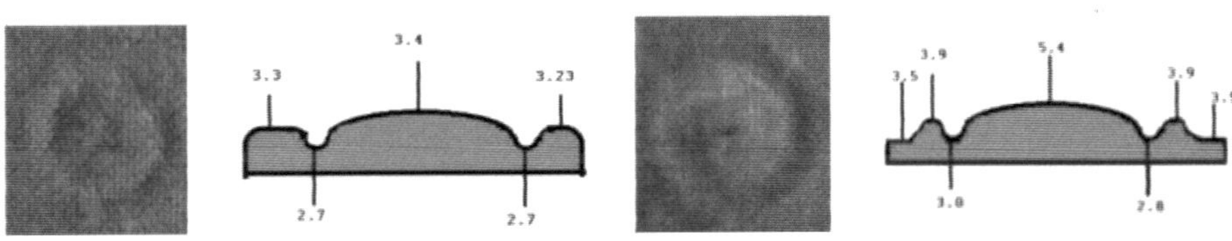

Figure 7: a) Thickness variation at vacuum supply b) Thickness variation at resin supply

Mechanical comparator with dial gauge was used to measure laminate thickness at 25 different locations. The readings were taken randomly at different location for each laminate. It was observed that the thickness at vacuum supply and resin supply location was different than at other locations. Refer figure 7 (a) and (b) to study the variation in thickness at resin supply location and vacuum supply location.

Except covering the areas shown above, other readings were taken at 25 locations, randomly on full laminate. Figure 8 shows maximum variation in thickness and mean thickness of each laminate. The minimum variation of 0.17 mm in four layers and maximum variation of 1.3 mm in six layers has been observed over the entire thickness of laminate. This shows that as we increase number of layers the variation in thickness also increases.

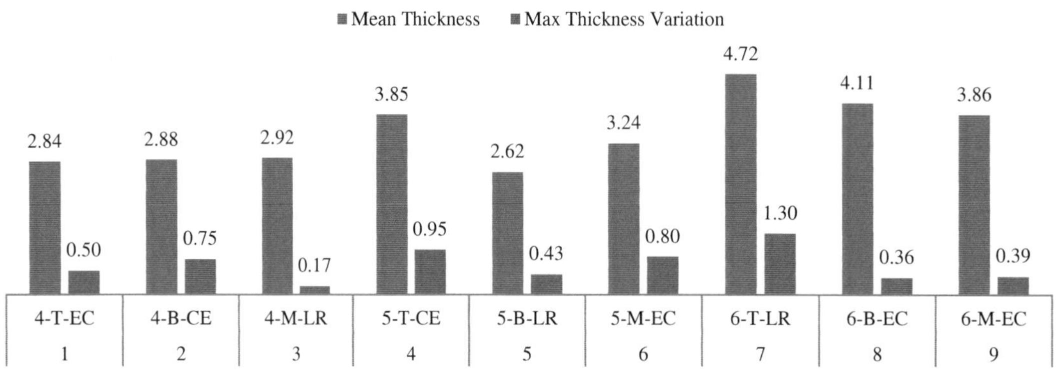

Figure 8: Thickness variations in laminates.

5.3 Fibre weight fraction

To find fibre weight fraction solvent method is been used by authors. Tetra Hydro Furan (THF) is been used as solvent. The laminates were cut in size of 10 mm X 10 mm and initial weights were noted after which the laminates were kept in solvent for 48 hours and again the weight was measured after removal of resin. As per this, 70% fibre weight fraction was observed. To find out fibre volume fraction rule of mixture can be applied.

6. Conclusion

The work was performed to investigate the importance of position of resin supply, vacuum supply location and number of layers in making laminate of jute- polyester FRP composite with VARTM process. The maximum tensile strength observed when numbers of layers were 5; resin supply was from middle and vacuum supply was from edge to centre. As number of layers increased variation in laminate thickness increased. It has been found that by increasing flow velocity tensile strength did not increase. Solvent method was used to find fibre weight fraction of jute in laminate and 70% fibre weight fraction was achieved average in all laminates.

References

[1] Williams C, Summerscales J and Grove S 1996 Resin infusion under flexible tooling (RIFT): a review *Compos. Part A Appl. Sci.,* **27(7),** 517-24

[2] Summerscales J and Searle T J 2005 Low-pressure (vacuum infusion) techniques for moulding large composite structures *P I MECH ENG L-J MAT,* **219(1),** 45-58

[3] Bhatt A T, Gohil P P and Chaudhary V 2018 Primary manufacturing processes for fiber reinforced composites: History, development & future research trends *Mater. Sci. Eng.,* **330,** 012107

[4] Kuentzer N, Simacek P, Advani S G and Shawn W 2007 Correlation of void distribution to VARTM manufacturing technique *Compos. Part A Appl. Sci.,* **38(3)**, 802-13

[5] Kedari V R, Farah B I and Hsiao K 2011 Effects of vacuum pressure, inlet pressure, and mold temperature on the void content, volume fraction of polyester/ e-glass fiber composites manufactured with VARTM process. *J. Compos. Mater.*, **0(0)**, 1-16

[6] Rodriguez E S, Francucci G and V Analia 2008 Study of saturated and unsaturated permeability in natural fiber fabric *The 9th International conference on flow process in composite materials.*

[7] Francucci G, Rodriguez E S and Analia V 2010 Study of saturated and unsaturated permeability in natural fiber fabrics *Compos. Part A: Appl. Sci. and Manuf.*, **41(1)** 16-21

[8] Tripathi A and Shukla M 2015 Determination of permeability of polymer matrix composites produced by VARTM *IJAPME,* **1(6),** 6202-10

[9] Grimsley B W, Hubert P, Song X L, Cano R J, Loos A C and Pipes R B 2001 Flow and compaction during the vacuum assisted resin transfer molding process.

[10] Gama B A, Li H, Li W, Paesano A, Heider D and Gillespie J W 2001 Improvement of dimensional tolerance during VARTM processing. *Proceedings of 33rd international SAMPE technical conference*. **33**, 1415–27
[11] Yoon M K, Baidoo J, Gilliespie J W and Heider D 2005 Vacuum assisted resin transfer moulding (VARTM) process incorporating gravitational effects: A closed-form solution. *J. Compos. Mater.*, **39(24)**, 2227- 42
[12] Rigas E J, Mulkern J T, Walsh S M and Nguyen S P 2001 Effects of processing conditions on vacuum assisted resin transfer moulding process VARTM." *ARL-TR-2480.*
[13] Heider D and Gillespie W 2010 VARTM variability and substantiation. *Proc. Jt. Adv. Mater. Struct. Cent. Excell.*
[14] Ghabezi P, Golzar M and Jalall S R 2010 Effect of inlet tube dimensions on VARTM mould filling time. *The 2nd International Conference on Composites: Characterization, Fabrication and Application.*
[15] Chang C and Chen W 2016 Experimental study of progressive compression method of resin deliver in liquid compression moulding.*Proc. eng. technol. innov.*, **2**, 23-25
[16] Plunkett B 2010 Vacuum Assisted Resin Transfer Molding (VARTM) System. *Advanced Materials and Technologies Laboratory Mechanical Engineering. Department, Virginia Tech Summer.*
[17] Ho M, Wang H, Lee J H, Ho C, Lau K, Leng J and Hui D 2012 Critical factors on manufacturing processes of natural fiber composites. *Compos. B. Eng.*, **43(8)**, 3549-62
[18] Sanjay M R, Arpitha G R, Naik L, Gopalakrishna K and Yogesha B 2016 Applications of natural fibers and its composites: An overview. *Compos. B. Eng.*, **7(3)**, 108
[19] Deb A, Das S, Mache A and Laishram R 2017 A study on the mechanical behaviors of jute-polyester composites. *Proceedia Engineering* **173** 631-38
[20] Das S and Bhowmick M 2015 Mechanical properties of unidirectional Jute- Polyester composite. *J Textile,* **5(4)**, 1
[21] Zikre H and Bhatt A T 2016 Comparison of mechanical properties of fiber reinforced plastic laminates composite with different thickness, manufacturing techniques and structure. *Int Conf Multidiscip Res Pract.*, **4(1)**, 62e7
[22] Babu B Z and Pillai K M 2004 Experimental investigation of the effect of fiber-mat architecture on the unsaturated flow in liquid composite moulding. *J. Compos. Mater.*, **38(1)**, 57-79
[23] Anil, K C, Girisha K G and Sreenivas Rao K V 2015 Effect of fiber orientation on specific gravity, hardness and flexural strength and tensile property of jute/hemp hybrid laminate composite. *Appl. Mech. Mater.*, **766**, 75-78

IOP Conf. Series: Materials Science and Engineering 1004 (2020) 012002 doi:10.1088/1757-899X/1004/1/012002

Patch shape and size effect on performance of in-situ scarf patch repair in aircraft panel

Kishan Dwivedi [1] and Himanshu Pathak[1*]

[1]School of Engineering, Indian Institute of Technology Mandi, Mandi, Himachal Pradesh-175005, India.

[*]Email: himanshu@iitmandi.ac.in

Abstract. Patch repairing technique is effective and efficient approach to enhance structural performance of aged aircraft panels. The presented work emphases on studying the patch shape and size effect on scarf composite patch repair of aluminium aircraft panel under thermal, thermo-elastic and mechanical load. In this work, patched repaired structural member has been simulated by ABAQUS software to find out the performance of patch repair in term of patch thickness and patch shape. Five different patch shapes are chosen to analysis: circular, square, hexagonal, rectangular and trapezoidal with different patch thickness 1, 2, 3, 4, 5 mm. The Heat transfer and coupled temp-displacement analysis has been performed under cure temperature 394 K for in-situ patch repair model. The analysis of the critical von-Mises stress and heat flux load distribution in composite patch repair system allows us to estimate performance in composite patch repair system. The results show that patch shape and size has significant effect on magnitude of heat flux load and induced von-Mises stress. Trapezoidal shape of composite patch repair has lower value of von-Mises stress and heat flux load as compared to all four shapes of composite patches.

1. Introduction

Structural members used in aircraft application subjected aerodynamic loading in service condition. These loads can weaken the structural member and reduces the load carrying capacity. The out-dated way to deal with such problems involves mechanical fasteners with metallic patches. The composite patch repair have several benefits over mechanical fastened repairing method, such as strength and life of patch repair system is improved, obstacle in crack propagation [1]. For best performance of a bonded repair, it is essential to fully understand the thermal response of these repairs, which can in future improve their processing, and design of repair methods and guiding principle for composite structures [2]. Difference in thermal expansion coefficient (alpha) at the contact surface of patch and panel is responsible for create the thermal stress during curing process so that studies of curing process is become so important for effective patch repair. [3]. Patch thickness is one of the important parameter for studies of one side composite patch repair [4]. Rachid et. al. [5] studies different shape of patch repair like rectangular shape, H shape, arrow shape and analyse the effect of repair durability and efficiency on patch repair of aircraft structure by using three dimensional finite element method. He reported that the repair efficiency of arrow shape of patch repair is higher then compare to rectangular and H shape of patch repair. Benyahia et. al. [6] studied circular, elliptical shape of patch repair and make a comparison between them through stress intensity factor analysis at crack tip under thermal and mechanical load by using 3D finite element method. It has been found that circular shape of patch repair can improve the efficiency of repair because this shape have less value of stress intensity factor. Kashfuddoja et. al. [7]

IOP Conf. Series: Materials Science and Engineering 1004 (2020) 012002 doi:10.1088/1757-899X/1004/1/012002

studied for oval, ellipse, circular, octagonal, square, rectangle shape of patch repair and estimate the effect of patch shape on efficiency of repair by using 3D finite element method. He analyse the peal stress and stress concentration factor for finding the optimum shape of patch repair. T.D. Breitzman et. al. [8] studied for optimum of scarf composite patch repair in tensile load condition by using 3D nonlinear analysis. The presented work predicted distribution of stress over adhesive and conclude for efficient composite repair design. Tie Y. et. al. [9] studied for impact analysis of square, rhombus, circle, hexagon, octagon shape of patch repair by using experimental and numerical method. Optimal impact resistance were presented for patch repair model. It was found that circular patch lead to better impact resistance behaviour.

From the literature discussion, it can be concluded that both thermal and mechanical analyses were done to predict structural performance of patch repairing model. However, in-situ patch repair problem is not much more explored by scientific community. The present work analyses performance of in-situ patch repairing problems under thermal and mechanical loading environment. Following objectives laid down for the presented work:

- Heat transfer, coupled temp-displacement and mechanical load analyses have been presented using three dimensional finite element method.
- Performance prediction of square, circular, hexagonal, rectangular and trapezoidal shape patch repair.
- Numerical results are presented in the form of induced von-Mises stress and heat flux.

2. Geometrical and materials models

Consider an elastic repaired panel is made of aluminum material have following dimensions: width 30 mm, length 30 mm and thickness 5 mm. This aluminum panel is repaired with circular, square, hexagonal and rectangular shape of patch repair have boron / epoxy material. Heating blanket in patch repair system is made of silicon material have following dimension like width 10 mm, length 10 mm and thickness 1 mm. The following dimension of different shape of composite patch is mentioned bellow.

- Circular composite patch: Radius 5 mm and patch thickness is varies like 1, 2, 3, 4, 5 mm.
- Square composite patch: length 10 mm, width 10 mm and thickness is varies like 1, 2, 3, 4, 5 mm.
- Hexagonal composite patch: max length 10 mm, max width 10 mm and thickness is varies like 1, 2, 3, 4, 5 mm.
- Rectangular composite patch: length 10 mm, width 4 mm and thickness varies like 1, 2, 3, 4, 5 mm.

Trapezoidal composite patch have dimension of two parallel side is 20 mm, 10 mm, and both tapered side length is 7.07 mm and width 10 mm, thickness 5 mm. The dimension of aluminum panel for trapezoidal composite patch repair is given as 15 mm, 10 mm for parallel side, width 10 mm and thickness 5 mm. Heating blanket have following dimension like width 10 mm, length 20 mm, thickness 2.5 mm for trapezoidal composite patch repair. Heating blanket is subjected to cure temp 394 K, and it is assumed that material like aluminum, composite patch and silicon have the elastic and isotropic property. The elastic property of patch, panel, heating blanket is given in table 1.

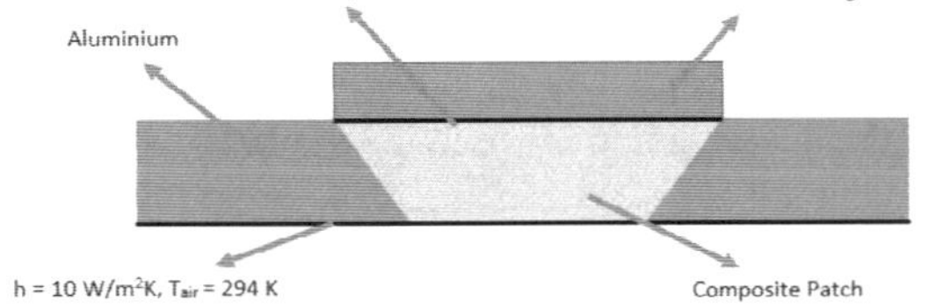

Figure 1: Trapezoidal composite patch repair.

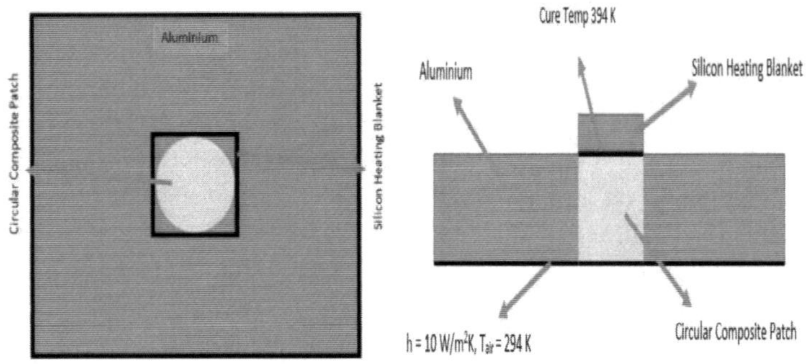

(a) (b)

Figure 2: Circular composite patch repair (a) Top view (b) Side view.

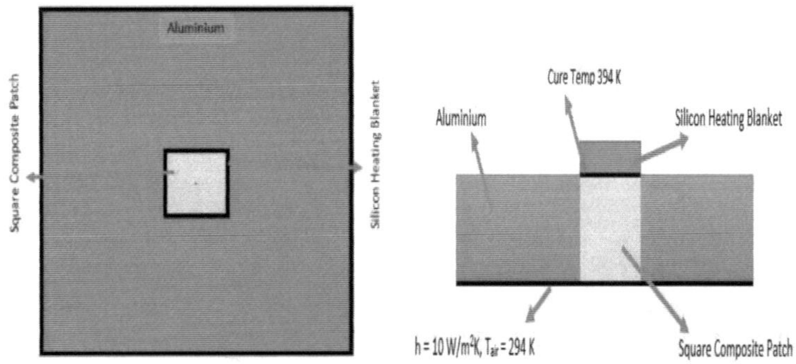

(a) (b)

Figure 3: Square composite patch repair (a) Top view (b) Side view.

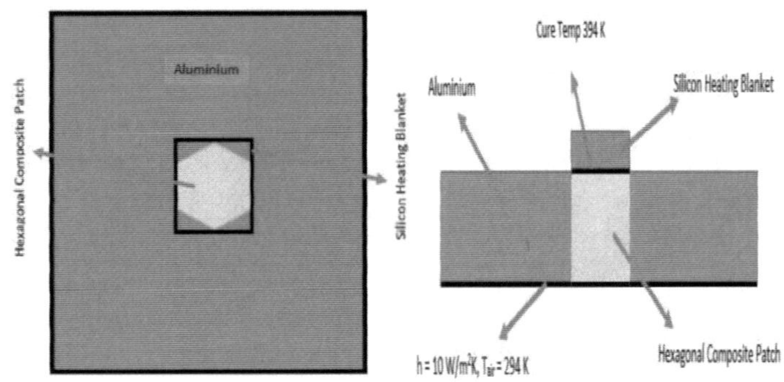

(a) (b)

Figure 4: Hexagonal composite patch repair (a) Top view (b) Side view.

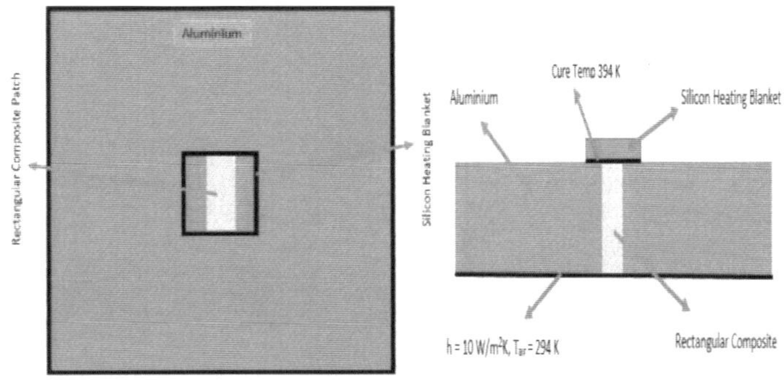

(a) (b)

Figure 5: Rectangular composite patch repair (a) Top view (b) Side view.

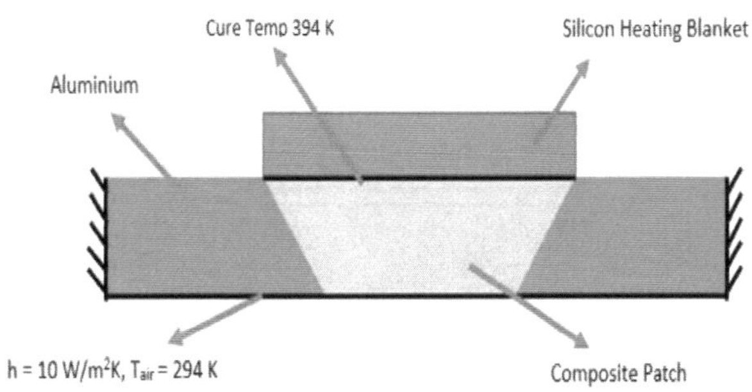

Figure 6: Trapezoidal composite patch repair with both ends are fixed.

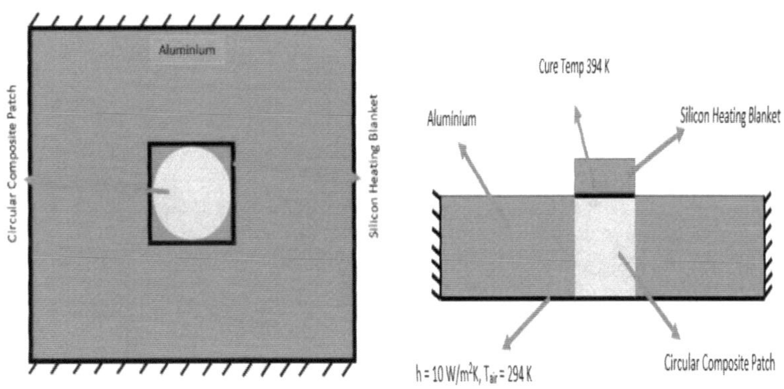

(a) (b)

Figure 7: Circular composite patch repair with fixed end (a) Top view (b) Side view.

IOP Conf. Series: Materials Science and Engineering 1004 (2020) 012002 doi:10.1088/1757-899X/1004/1/012002

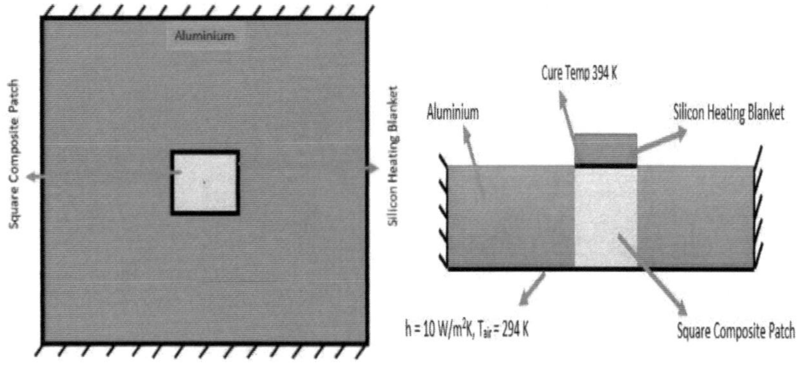

(a) (b)

Figure 8: Square composite patch repair with fixed ends (a) Top view (b) Side view.

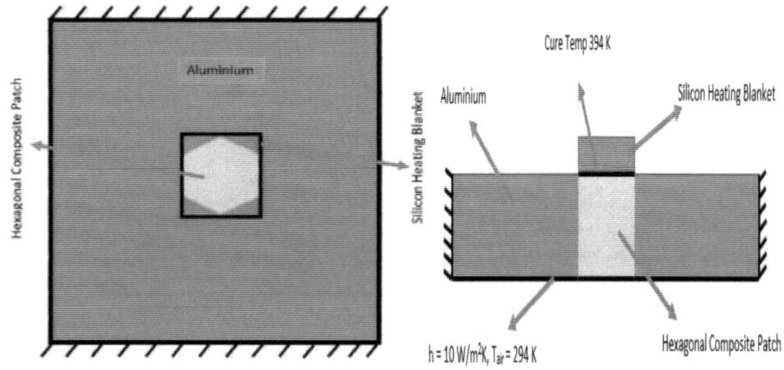

(a) (b)

Figure 9: Hexagonal composite patch repair with fixed end (a) Top view (b) Side view.

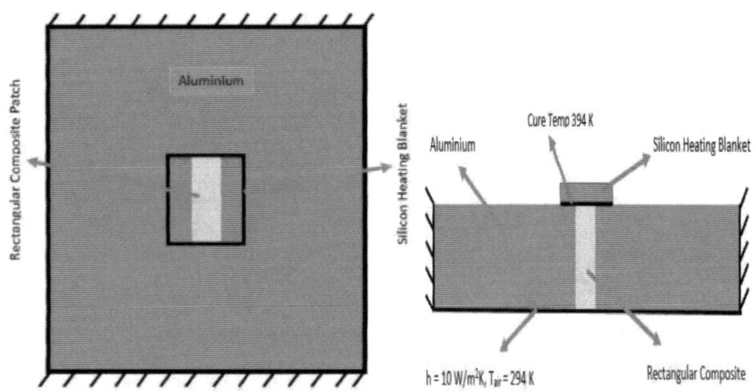

(a) (b)

Figure 10: Rectangular composite patch repair with fixed ends (a) Top view (b) Side view.

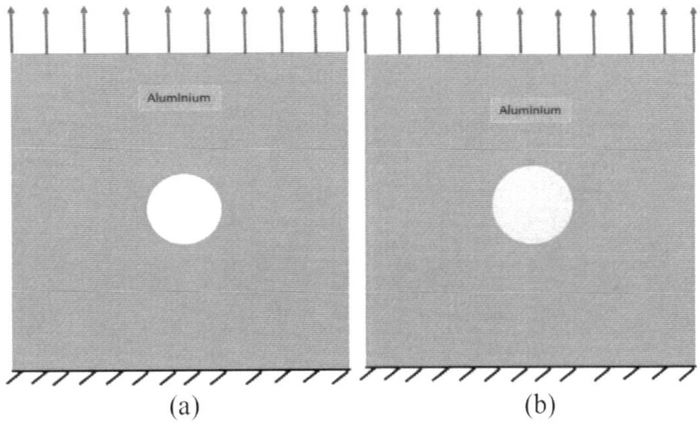

Figure 11: Top view of circular (a) unpached repair (b) patched repaired, fixed at one end and have tensile traction load on other end.

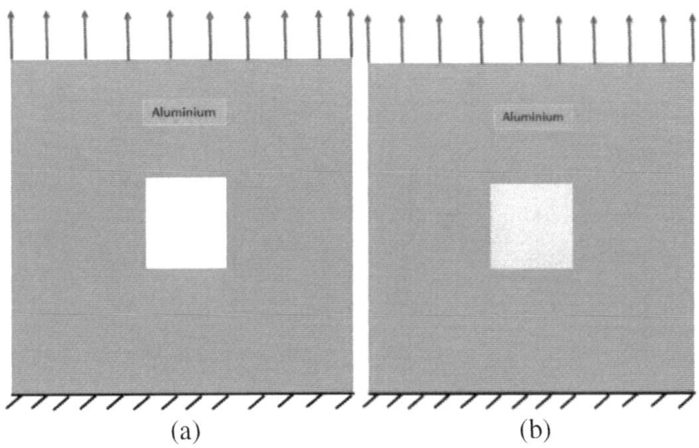

Figure 12: Top view of square (a) unpached repair (b) patched repair, fixed at one end and have tensile traction load on other end.

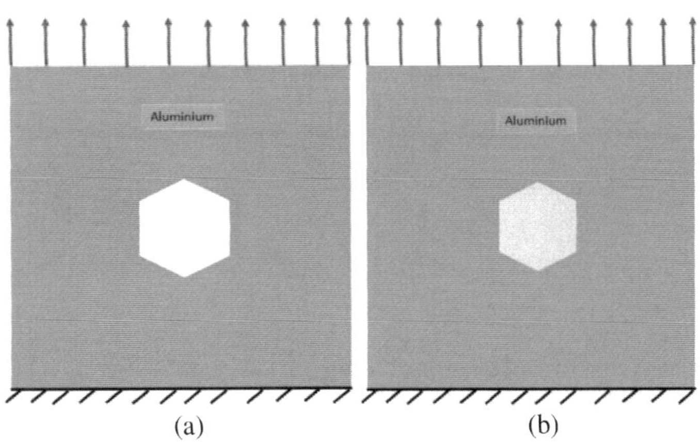

Figure 13: Top view of hexagonal (a) unpached repair (b) patched repair, fixed at one end and have tensile traction load on other end.

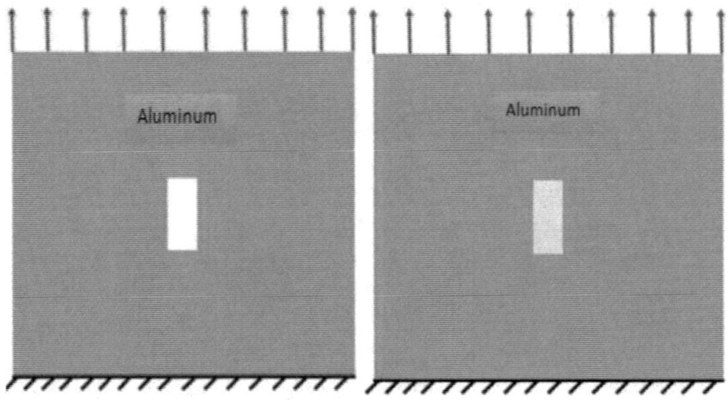

Figure 14: Top view of rectangular (a) unpached repair (b) patched repair, fixed at one end and have tensile traction load on other end.

3. Finite element modelling

Three dimensional finite element analysis is carried out in this current work by using ABAQUS software. The finite element model consist of three subsections like, composite patch, aluminum panel and silicon heating blanket. The aluminum plate is mesh with three solid element in the direction of thickness and the heating blanket is mesh with one solid element in direction of thickness, and composite patch are mesh with 10, 8, 6, 4, 2 solid element in thickness direction respectively for 5, 4, 3, 2, 1 mm thickness for all shape of composite patch. Aluminum panel have 1617, 1680, 2160, 2160 number of element for circular, square, hexagonal, rectangular shape of composite patch repair, silicon heating blanket have 100 number of element for circular, rectangular, square, hexagonal shape of composite patch repair model. The circular composite patch have 768, 1536, 2304, 3072, 3840 number of element for 1, 2, 3, 4, 5 mm thickness of patch respectively. The square composite patch have 800, 1600, 2400, 3200, 4000 number of element for 1, 2, 3, 4, 5 mm thickness of patch respectively. The hexagonal composite patch have 320, 640, 960, 1280, 1600 number of element for 1, 2, 3, 4, 5 mm thickness of patch respectively. The rectangular composite patch have 320, 640, 960, 1280, 1600 number of element for 1, 2, 3, 4, 5 mm thickness of patch respectively. Trapezoidal patch repair model have 165, 2940, 50 number element of aluminum panel, composite patch and silicon heating blanket respectively.

For heat transfer simulation, model (aluminum panel, composite patch, silicon heating blanket) have DC3D8 hexahedral brick element. For coupled temp-displacement simulation, model have C3D8T hexahedral brick element and for mechanical load simulation, model have C3D8R hexahedral brick element.

Simulation have steady state analysis for three different type of "STEP" option like heat transfer, couple temp-displacement and static general. This "STEP" option is used in ABAQUS simulation. The total time period was given one second with maximum number of increment is 100, minimum increment size was given as 1×10^{-5} and maximum increment size is one.

Heat transfer and couple temp-displacement analysis of circular, square, hexagonal, rectangular, trapezoidal shape of patch repair in ABAQUS, conducted as follows: material property of aluminum, composite patch, silicon is given as shown in the table 1. Heat transfer is chosen as "STEP" option for heat transfer analysis and for couple temp-displacement is chosen as "STEP" option for couple temp-displacement analysis in ABAQUS. The Cure temp of 394 K was set to surface of the heating blanket and surface film condition as a boundary condition is given on back side of aluminum and composite patch as shown in the figure number from 1 to 10, these both boundary condition is set for both heat transfer analysis and couple temp-displacement analysis. Couple temp-displacement analysis have one extra boundaries condition in which both opposite end of the aluminum panel is set to "ENCASTER" as shown in the figure number from 6 to 10 for all four shape of composite patch repair.

Mechanical load analysis was carried out for circular, square, hexagonal, rectangular shape of patch/unpatched repair in ABAQUS. Material property is chosen from table 1 and static general is chosen for "STEP" option in ABAQUS. One end of the aluminum panel is given "ENCASTER" boundary condition and in opposite end uniform tensile traction load is applied of magnitude 10, 20, 30, 40, 50 GPa as shown in the figure number from 11 to 14.

Table 1: Materials properties (panel, patch and heating blanket).

Property, (SI Unit)	**Panel**	**Patch**	**Blanket**
Thermal Conductivity, (W/m.K)	205	0.21	150
Density, (Kg/m^3)	2710	1310	2330
Specific Heat Capacity, (J/kg.K)	921.1	1260	710
Thermal expansion Coefficient, (10^{-6} °C^{-1})	23	43	2.6
Elastic Modulus, (GPa)	700	386	1790
Poison ratio	0.3	0.3	0.3

4. Result and Discussion

This current work is carried out for to find the best patch shape and size, which can improve the performance of patch repair. The analysis of heat flux and von-Mises stress distribution over the composite patch repair gives a way to find this performance. The five patch shape like circular, rectangular, hexagonal, square and trapezoidal are compared to each other based on their performance.

4.1. Thermal loading

In this section, the variation of the heat flux load over the circular, rectangular, hexagonal, square shape of composite patch repair are compare with different patch thickness for heat transfer analysis as shown in the figure 15. One can note, from this figure 15 heat flux load is decreasing with increase in patch thickness for each shape of composite patch repair. It can also note, in this figure 15 square shape of patch present the lover value of heat flux load and rectangular shape of patch present higher value of heat flux load for all patch thickness. The circular and hexagonal shape of composite patch repair have approximately same value of heat flux load for patch thickness 2, 3, 4, 5 mm and have small difference in value heat flux load for patch thickness 1 mm. The trapezoidal shape of composite patch repair have value of heat flux load is 15310 W/m^2 in heat transfer analysis, this value of heat flux load is very small then compare the above mentioned all four shapes of composite patch repair. The line contour and legend of heat flux load in heat transfer analysis for circular composite patch repair of thickness 5 mm is present in figure 16.

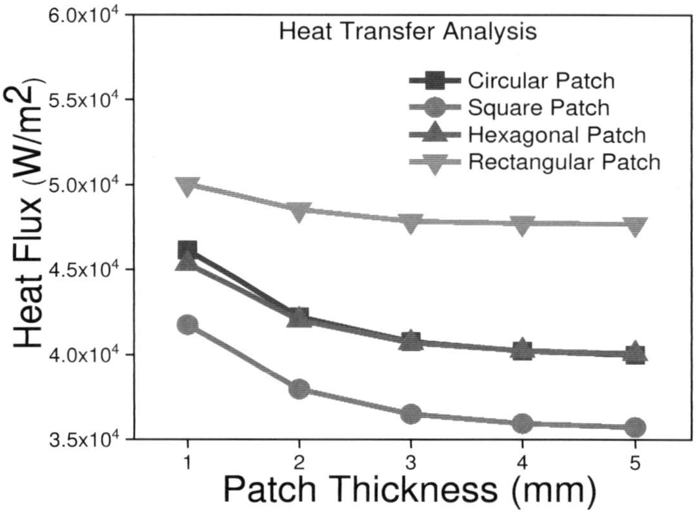

Figure 15: variation of heat flux load with composite patch thickness in heat transfer analysis.

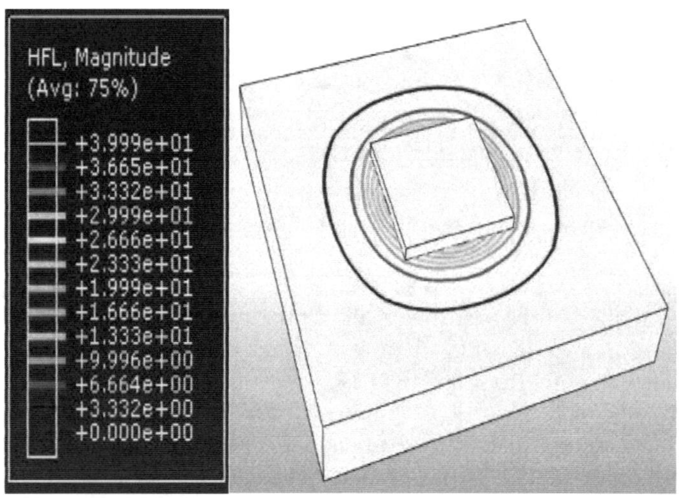

Figure 16: Line contour of heat flux load (10^{-3}W/mm^2) of circular composite patch repair model of thickness 5 mm in heat transfer analysis.

4.2. Thermo elastic loading

In this section, the variation of induced von-Mises stress and heat flux are consider for circular, rectangular, square, hexagonal shape of composite patch repair and compare with different patch thickness for couple temp-displacement analysis as shown in the figure (17,19). One can note, from the figure 17 rectangular shape of composite patch repair present lower magnitude of von-Mises for all patch thickness, as compared to all four shape of composite patch repair. It can also note that up to 4 mm patch thickness the von-Mises stress is higher for hexagonal shape of composite patch repair but for the 5 mm patch thickness circular shape of patch repair have higher value in compare of all four shape of patch repair. Square shape of patch repair have von-Mises stress lower then compare hexagonal and circular shape of patch repair but have higher value then compare to rectangular shape of patch repair. The value of the von-Mises stress for trapezoidal shape of patch repair is $1.45x10^{10}$ N/m^2 which is lower then compare to all shape of composite patch repair. Line contour of von-Mises stress for circular patch repair of thickness 5 mm in couple temp-displacement analysis as shown in figure 18.

From the figure 19 it can be noted, that the value of heat flux is increasing with composite patch thickness for all the four shape of patch repair. The circular shape of patch repair have higher value of heat flux load and square shape of patch repair have lover value of heat flux load then compare to all four shape of patch repair. The trapezoidal shape of composite have the value of heat flux is $2x10^4$ W/m^2 which is too small then compare to heat flux magnitude of all other four shape of patch repair in couple temp-displacement analysis. Line contour of the heat flux load of circular patch repair of thickness 5 mm in couple temp-displacement analysis is present in figure 20.

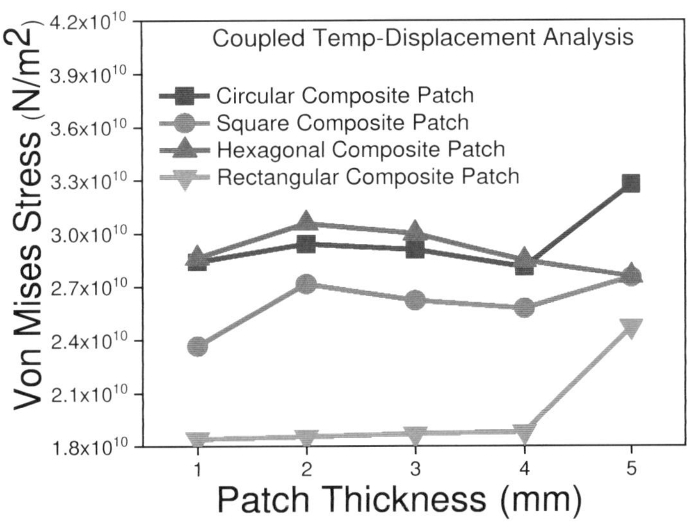

Figure 17: Von-Mises stress variation with composite patch thickness in couple temp-displacement analysis.

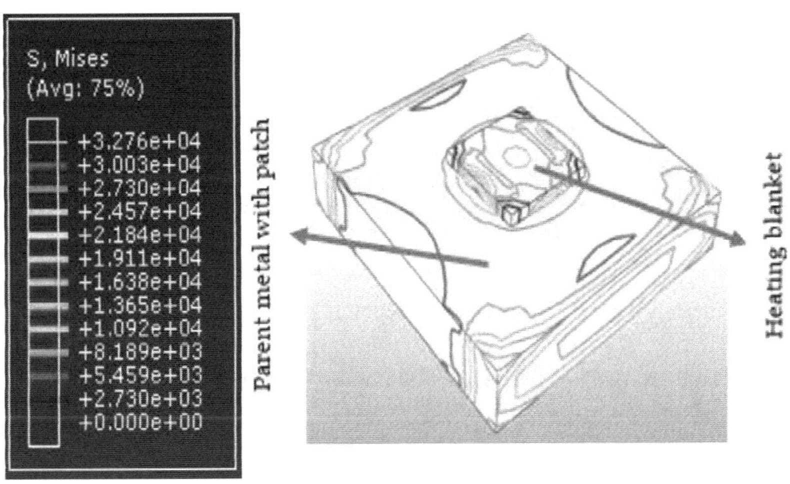

Figure 18: Line contour of von-Mises stress (N/mm^2) in circular composite patch repair model of thickness 5 mm in couple temp-displacement analysis.

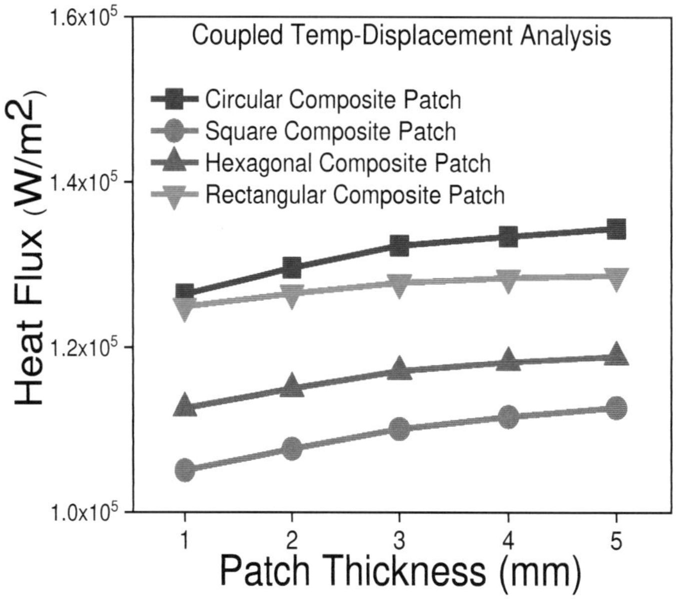

Figure 19: Heat flux variation with composite patch repair thickness in couple temp-displacement analysis.

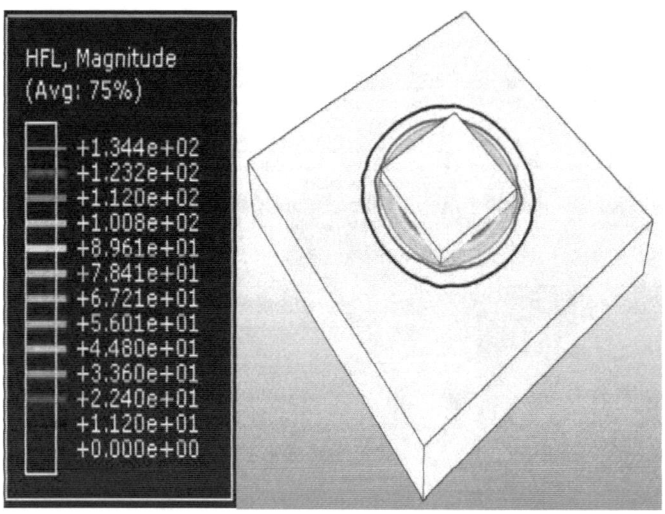

Figure 20: Line contour of heat flux (10^{-3}W/mm^2) of circular composite patch repair model of thickness 5 mm in couple temp-displacement analysis.

4.3. Mechanical loading

In this section the variation of induced von-Mises stress is consider for circular, rectangular, square, hexagonal shape of composite patch/unpatched repair of thickness 5 mm and compare with tensile traction load (10, 20, 30, 40, 50 GPa) for mechanical load analysis as shown in the figure 21. From figure 21, induced von-Misses stress is increasing with magnitude of applied tensile traction load.

The circular shape of patch/unpatched repair model shows higher value of von-Mises stress then compare to all shape of patch/unpatched repair respectively for each tensile traction load. The rectangular shape of patch/unpatched repair model shows lower magnitude of von-Mises stress then compare to all shape of patch/unpatched repair for each tensile traction load. Line contour of von-Mises stress for circular patch repair of thickness 5mm for mechanical load analysis is present as shown in figure 22.

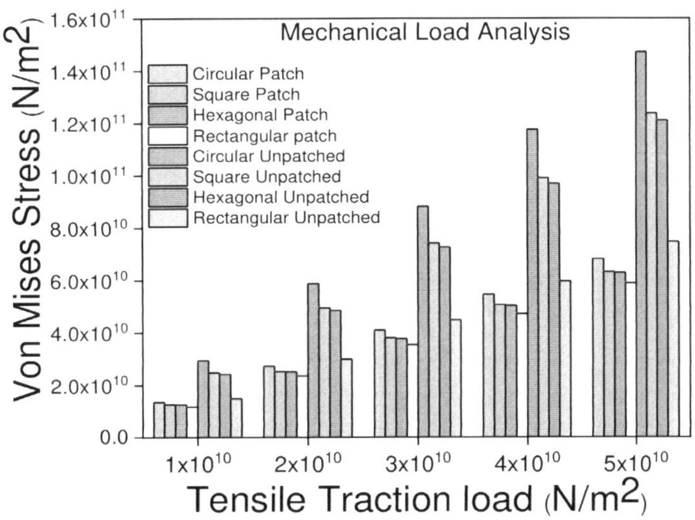

Figure 21: Von-Mises stress variation with tensile traction load in mechanical load analysis.

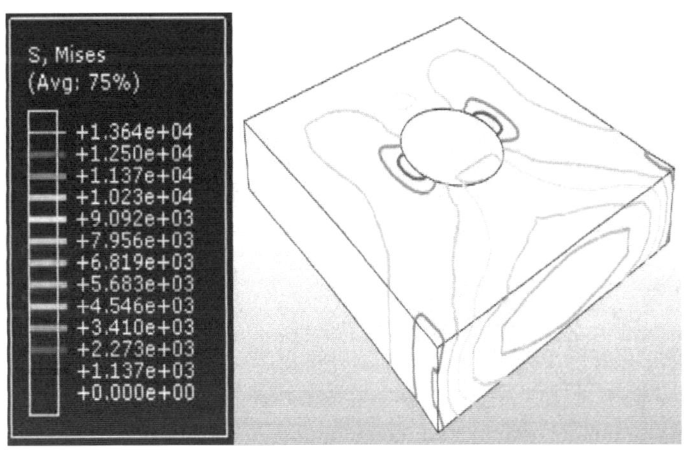

Figure 22: Line contour of von-Mises stress (N/mm^2) for circular composite patch repair of thickness 5 mm in mechanical load analysis.

5. Conclusions

The current work has main aim to compare the performance of square, rectangular, hexagonal and circular shape of patch repair with trapezoidal shape patch repair under the thermal loading, thermo-elastic loading and mechanical loading. Three-dimensional finite element model has been implemented to perform numerical analyses using ABAQUS package. The trapezoidal shape of patch repair have lower value of heat flux and induced von-Mises stress then compare to all four shape of patch repair in all type of analysis. Trapezoidal shape of patch repair can improve the performance then compare to all four shape of patch repair. Following conclusions can be drawn based on presented numerical investigations:

- Heat transfer analysis result shows least value of heat flux for square composite patch, which improve the repair performance compare to all four shape of composite patch repair.
- Coupled temp-displacement analysis results shows the lower magnitude of von-Mises stress and heat flux load for rectangular and square shape of composite patch repair respectively.
- Mechanical load analysis result shows rectangular shape of composite patch/unpatched repair have least magnitude of von-Mises stress with respect to each tensile traction load, which can improve the repair performance.
- Patch thickness has significant effect on repair performance.

Acknowledgement

Financial support received from Aeronautics Research & Development Board (DRDO), New Delhi, India through grant file no. 1051914 is gratefully acknowledged.

References

[1]Kumar A M and Hakeem S A 2000 Optimum design of symmetric composite patch repair to centre cracked metallic sheet *J. of Composite structures*., **49**, 285-292.

[2]http://digitool.library.mcgill.ca/webclient/StreamGate?folder_id=0&dvs=1578389405604~85, Visited, 15th November 2019.

[3]Naboulsi S and Mall S 1997 Thermal effects on adhesively bonded composite repair of cracked aluminum panels *J. of theoretical and applied fracture mechanics*., **26**, 1-12.

[4]Khan S M A and Essaheb M 2017 Effect of patch thickness on the repair performance of bonded composite repair in cracked aluminium plate *J. of materials today: proceedings*., **04**, 9020-28.

[5]Rachid M, Serier B, Bouiadjra B B and Belhouari M 2012 Numerical analysis of the patch shape effects on the performances of bonded composite repair in aircraft *J. of Composites: Part B*., **43**, 391-397.

[6]Benyahia F, Albedah A and Bouiadjra B A B 2014 Elliptical an d circular bonded composite repair under mechanical and thermal loading in aircraft structures *Materials Research 2014*., ***54***, 18-24.

[7]Kashfuddoja M and Ramji M2014 Design of optimum patch shape and size for bonded repair on damaged carbon fiber reinforced polymer panels *J. of Matrials and Design*., **54**, 174-183.

[8]Breitzman T D, larve E V, Cook B M, Schoeppner G A and Lipton R P 2009 Optimization of composite scarf repair patch under tensile loading *J. of Composite: Part A*., **40**, 1921-30.

[9]Tie Y, Hou Y, Li C, Zhou X, Sapanathan T and Rachik M 2018 An insight into the low velocity impact behaviour of patch repaired CFRP laminates using numerical and experimental *J. of Composite structures*., **190**, 179-188.

Modeling and Simulation of a Semi-active Vehicle Suspension system using PID Controller

Prashant Narwade [1*], Ravindra Deshmukh [2], Mahesh Nagarkar[3] and Milind Bankar[4]

[1] Department of Mechanical Engineering, D.V.P.C.O.E, Ahmednagar, Maharashtra
[2] Department of Mechanical Engineering, J.N.E.C, Aurangabad, Maharashtra
[3]Department of Mechanical Engineering, S.C.M.C.O.E,, Ahmednagar, Maharashtra
[4]Tata Motors Ltd., Pune, Maharashtra
E-mail:*narwadeprashant@gmail.com

Abstract. In this article, the numerical simulation of Magneto-rheological (MR) damper has been done using PID controller in Matlab-Simulink software to compare between the regular Macpherson strut and semi active suspension systems. The Magneto-rheological damper is a control device filled with Magneto-rheological fluid which changes damping force by changing property of viscosity on application of magnetic field. The results show that proportional integral derivative (PID) controller used to control damper in semi-active suspension is more efficient to reduce road disturbances than the regular suspension using Macpherson strut. Vehicle chassis displacement and acceleration can be reduced within milliseconds of time.

Keywords. Magneto-rheological (MR) Fluid,, Sedimentation, Viscosity, Semi-active Suspension, Sprung mass

Nomenclature

m_1= Sprung mass, Kg
m_2= Unsprung mass, N/m
k_1= Suspension stiffness, N/m
k_2 = Tire stiffness, N/m
c_1= Damping coefficient of suspension, Ns/m
c_2= Damping coefficient of Tire, Ns/m
x_1= Displacement of the sprung mass, m
x_2= Displacement of the unsprung mass, m
w = Road disturbance, m
ξ = Damping factor
ω_n = Natural frequency of system, rad/s
ω = Forcing frequency of unsprung mass

1. Introduction

Suspension system along with tire is an important area of concern in a vehicle, which affects ride quality of a passenger car. A superior ride quality provides a comfort to passengers sitting in the vehicle. Speed bumps, pot holes and irregular road profile are main cause of vibrations in automobile. Roughness of road, rotating components coupled to the tire and engine creates vibrations during the ride of vehicle. Many researchers revealed that vibrations affect the human health and cause loss of working efficiency. Suspension system prevents these shocks and vibrations and provides comfort to the passengers. Along with comfort suspension system also carries the total load of the vehicle.

Damper is the most important element of the suspension assembly to ensure ride comfort and good control over the car by dissipating vibration energy. The automotive damper on most suspension systems are hydraulic and operate on the dashpot principle. Any relative motion between the axle and the car body results in a piston moving in a cylinder filled with oil. Considerable amount of viscous force oppose the relative motion across the car body, and this damping force is directly proportional to the velocity of vibrations. The most desirable amount of damping in these shock absorbers depends on the road condition. When running over a smooth road with rolling hills and valleys, critical damping is needed. On the other hand, if the road has short quick bumps, a small damping is desirable (J. P. Den Hartog, 1956[1]). However, it is appeared that the variability in the types of road unevenness is too great for changing the damping constant of passive damper. Adaptive suspension, Semi-active and active suspension can eliminate the drawback of regular suspension. To improve the ride quality and stability of vehicle, adaptive suspension can be used where it has a regular spring and adjustable damper with slow response. To increase the response time in milliseconds and control the damping force in real time, a semi-active suspension is used. Active suspension is a feedback controlled system that isolates the wheel from chassis of the vehicle body and effect of road roughness is not transferred to the passengers but at high cost. Semi-active suspension can also serve the purpose at low cost and can show reasonable performance as compared to that of an active suspension system. The rheological effect of controllable fluids, such as magnetorheological fluid or electro-rheological fluid is used in Semi-active system. Magneto-rheological (MR) fluids are rheological materials with magnetic particles mixed in carrier fluid along with additives or surfactants. Due to their variable rheological behaviour, Magneto-rheological fluids have bright future for applications in brakes, engine mounts and shock absorbers (Kciuk M. et al., 2009[3]).

2. Literature Review

To isolate the car body from disturbances of road, authors have studied on variable stiffness and variable damping methods. Yanqing Liu et al. have used two controllable dampers and springs, however controlling of spring is complicated to implement (Yanqing Liu et al. 2007[2]). Hence authors have tried changing damping force with element like magnetorheological damper. MR Damper is similar to regular hydraulic damper filled with fluid that moves between different chambers through small holes in the piston, with additional feature of an electrical coil around the piston which creates a magnetic field and changes the viscosity of oil. This change in viscosity is rapid and can come to original value within short period of time. By applying magnetic field in magnetorheological damper state of fluid can be changed from free flowing fluid to thick fluid like grease and vice versa. This concept can be used in different applications like suspension, MR Brake where braking torque is applied with the help of solid fluid and in magnetic bearings (Goncalves F. D et al., 2006[4]). High cost of MR fluids (US$ $816 per liter) and sedimentation problem is the major drawback of MR fluid (Sukhwani V. K and Hirani H, 2006[5]). Sedimentation problem can be solved by adding soluble surfactants or by using nano particles with magnetic properties and by coating particles with green additives .Nano particles or nano magnetized particles can improve sedimentation and stability but it also reduces the magnetic effect of the fluid (Bhau K. Kumbhar et al., 2015[6]). Strong nonlinear dynamical behaviour is exhibited by the magnetorheological damper which is a key challenge while solving equations. To characterize the dynamic behaviour of MR dampers, Bingham model, the classic Bouc-Wen model (Spencer et al., 1997[7]) and the modified Bingham hysteresis model (Yang et al., 2005[8]) are studied by the authors.

3. Analytical Modeling of a Quarter-Car Model

Two degree of freedom quarter car model is used to design a Suspension system. Figure 1 shows a chassis and vehicle body mass m_1 also known as sprung mass and mass m_2 known as unsprung mass which take account of suspension and tire. To generate equations of motion and solve numerical simulations for vibration isolation performance, free body diagram of one of the four wheels (Quarter car model) is presented.

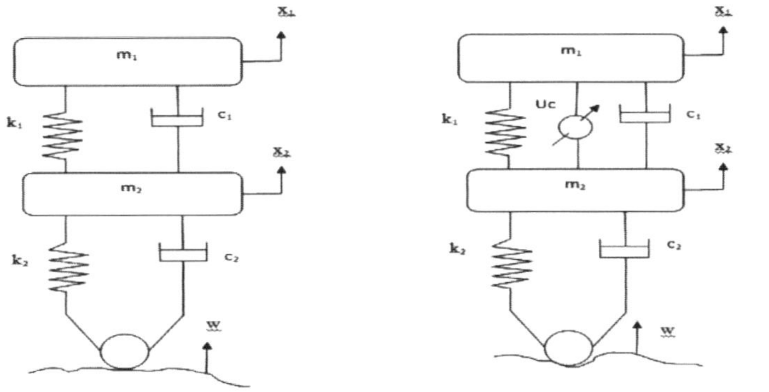

a) Passive Suspension system b) Semi-active Suspension System

Figure 1: Free body diagram of quarter car model

Vehicle comfort and ride quality depends on the acceleration of vibration transferred by the sprung mass to the end user. Lower the acceleration, better is the comfort and with high acceleration vibration increases. The equation developed from Newton's second law for acceleration of sprung mass is given as follows

$$\ddot{x}_1 = \omega^2 \,.\, x_1 \,.\, \sqrt{\frac{1+(2\xi r)^2}{(1-r^2)^2+(2\xi r)^2}} \qquad \textbf{(1)}$$

Where ξ is damping factor, ω_n is the natural frequency of system, ω is the forcing frequency of unsprung mass excitation, k_1 is the sprung mass stiffness, k_2 is stiffness of tire, r is the frequency ratio ($\omega/\ \omega_n$). From equation 1 it is observed that the mass of vehicle and forcing frequency affects the acceleration of vibration. .The equations of motion for the vibratory system modeled are given as

$$m_1\ddot{x}_1 + c_1(\dot{x}_1 - \dot{x}_2) + k_1(x_1 - x_2) = 0 \qquad \textbf{(2)}$$

$$m_2\,\ddot{x}_2 - c_1(\dot{x}_1 - \dot{x}_2) - k_1(x_1 - x_2) + k_2(x_2 - w) + c_2(\dot{x}_2 - \dot{w}) = 0 \quad \textbf{(3)}$$

Where $\ddot{x}_1, \dot{x}_1, x_1,$ are acceleration, velocity and displacement of the sprung mass (quarter car body mass) and $\ddot{x}_2, \dot{x}_2, x_2,$ are acceleration, velocity and displacement of the unsprung mass respectively and $w, \dot{w}$ are road disturbance (displacement) and velocity of wheel due to road profile. The important parameter 'c_1' is the damping coefficient of sprung mass m_1, 'c_2' is the damping coefficient of unsprung mass m_2.The force generated by the controller is 'U_c' that takes into account the road profile and vertical displacement and velocity of the vehicle. In the model, PID controller is used in Matlab Simulink to correlate the control force (U_c) exerted by the controller. A Simulink model is build using the formulation from the equation (2) and (3) as shown in figure 2.

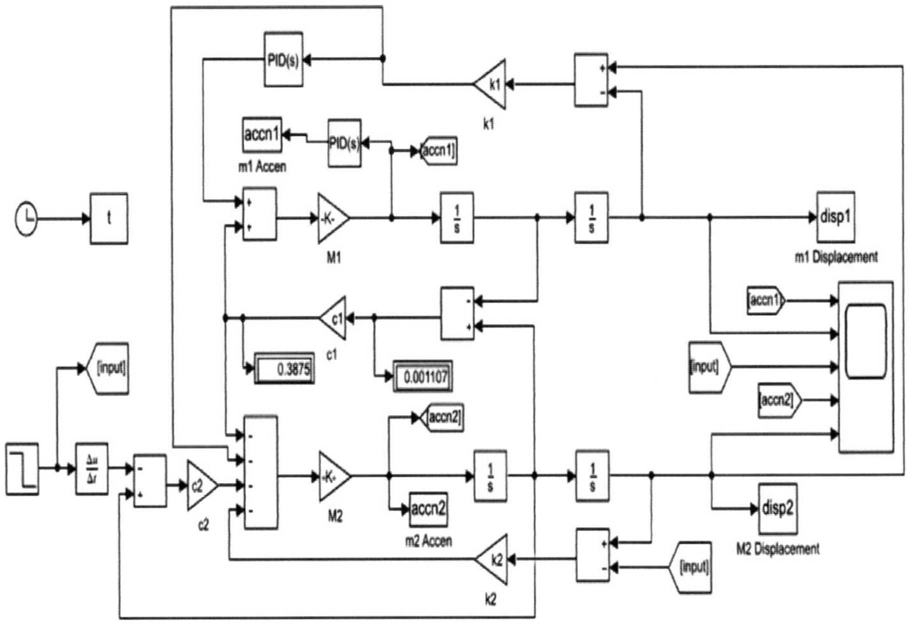

Figure 2: PID controller embedded in semi-active suspension system

To simulate and identify parameters of the suspension of car, PID controller is embedded in the semi-active suspension system. One PID controller is tuned for the displacement of the mass and other is tuned for the acceleration of the sprung mass.

4. Simulation Results and Discussions

The mathematical formulations generated from equations of motion are simulated in simulink model as shown in figure 2 with and without PID controller to compare the amplitude difference. Displacement and acceleration responses of the sprung mass with PID controller are compared with the displacement and acceleration values of a passively controlled system in suspension. The suspension parameters used in the quarter car model are shown in Table I.

Table 1: Suspension Element Parameters

Element Parameter	Value
m_1	240 Kg
m_2	45 Kg
k_1	23,000 N / m
k_2	1,98,000 N / m
c_1	450 N -s / m
c_2	0

During simulating model, tire damping coefficient is considered as zero, neglecting tire damping for finding the performance of the system using PID controller. Step input of 0.8 m is considered for the road disturbance ***'w'*** at tire. By solving the designed simulation, it is clearly observed that by defining the proportional, integral, derivative values, road disturbance at the sprung mass can be minimized. When vehicle with passive suspension is subjected to step input at wheels, acceleration obtained is as shown in figure 3, where it takes approximately 150 seconds to come to stable position. Also highest peak point of acceleration is observed at 140 m/s^2.But when PID controller is applied, vehicle comes to stable position in 50 seconds with maximum acceleration of 2 m/s^2 .Same concept of PID controller is correlated in MR damper to reduce the vibrations by controlling the damper. MR damper is controlled by supplying the current from 0.1 A to 2 A which subsequently changes the damping force. When system requires more time to settle or come to stable position, passengers in the car are subjected to vibrations for more period of time and exposure of vibration and noise for more period of time decreases the efficiency of human being. It also causes fatigue and discomfort to the passengers.

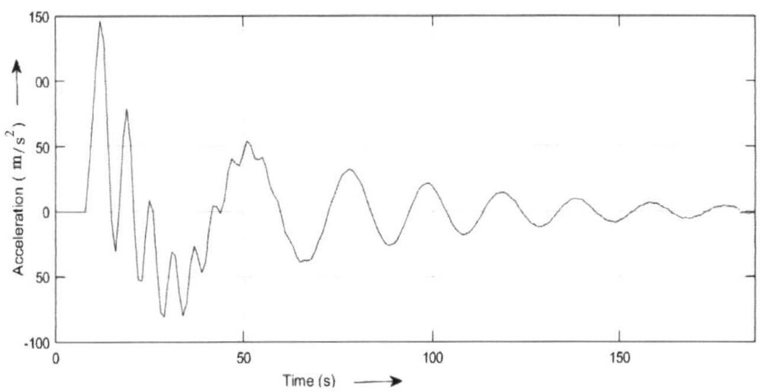

Figure 3: Acceleration of mass m_l without PID Controller

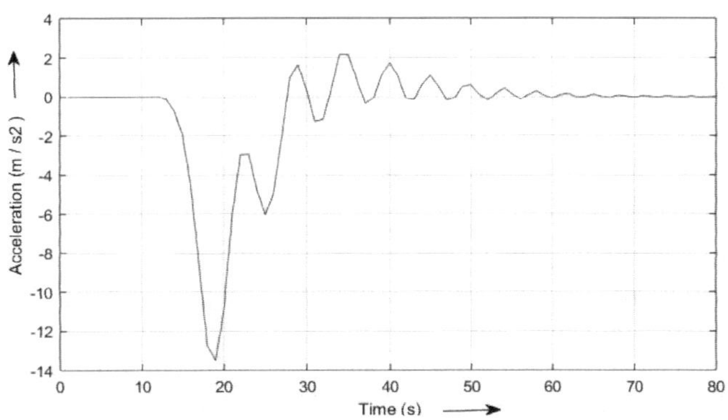

Figure 4: Acceleration of mass m_l with PID Controller

Same thing is observed in case of displacement of the vehicle with and without PID controller as shown in Figure 5 and Figure 6. It requires about 8 seconds for passive suspension system to become stable, where as PID controlled system come to rest in less than 2 seconds only for the step input applied. Sprung mass in passive suspension system is displaced for the step input from 0.8m to 1.4 m (0.6m), where as when PID controller is applied sprung mass is displaced from 0.8m to 1m (0.2m) only. Also passive suspension system behaves like under damped system, whereas PID control system looks like critically damped system coming to rest in short period of time.

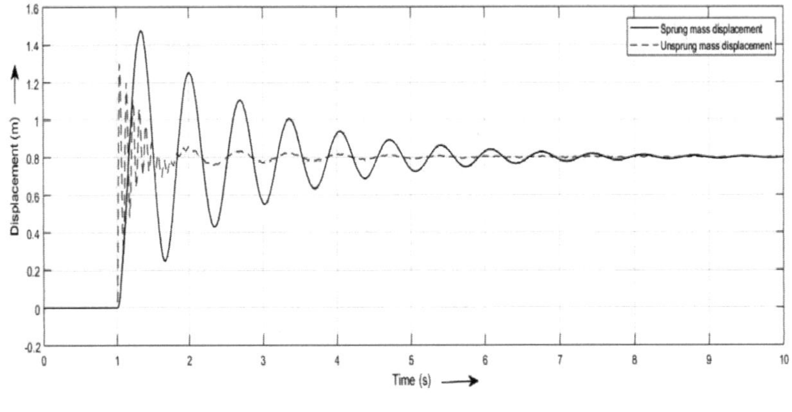

Figure 5: Displacement of m_1 without PID Controller

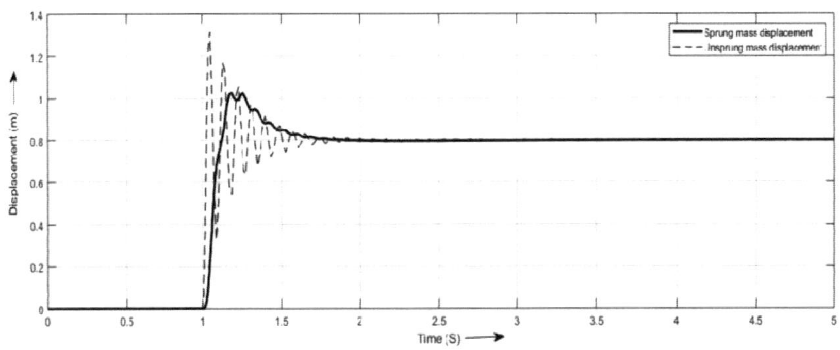

Figure 6: Displacement of m1 with PID Controller

5. Conclusions

The hysteresis behaviour of the MR fluid damper is formulated mathematically by developing quarter car model with PID controller in Simulink of MATLAB. The simulation result of PID controller have shown good results of vibration suppression over passively controlled damper model with the excitation signal of step input at wheels. With the PID controller, vehicle comes to stable position in 50 seconds with maximum acceleration of 2 m/s^2. Displacement of the system with PID controller comes to rest within short period of 2 seconds and behaves like critically damped system. This property of MR damper in suspension system provides comfort to the passengers.

Acknowledgement

The authors acknowledge Tata Motors Ltd., Pimri,Pune; Vijay kumar Sukhwani, Government Engineering college Ujjain; and Industrial Metal Powders, Pune for the support extended during this work.

References

[1] DenHartog J P 1956 Mechanical Vibrations *3rd ed., McGraw Hill Book Company Inc. New York,* 145-150.

[2] Yanqing Liu, Hiroshi Matsuhisa and Hideo Utsuno 2008 Semi-active vibration isolation system with variable stiffness and damping control *Journal of Sound and Vibration,* **313,** 16–28.
[3] Kciuk M, Kciuk S and Turczyn R 2009 Magnetorheological characterization of carbonyl iron based suspensions *Journal of achievements in materials and manufacturing engineering,* **33,** 135-141.
[4] Goncalves F D, Koo J H and Ahmadin M 2006 A review of the state of art in magnetorheological fluid technology – Part 1: MR Fluid and MR fluid models *Shock and Vibration Digest,* **38,** 203-219.
[5] Sukhwani V K and Hirani H 2006 Synthesis of a Magnetorheological lubricant *5th International conference on Industrial tribology ICIT-0, IISc Bangalore.*
[6] Bhau K Kumbhar , Satyajit R Patil and Suresh Sawant 2015 Synthesis and characterization of magneto-rheological (MR) fluids for MR brake application *Engineering Science and Technology,* **18,** 432-438.
[7] Spencer Jr. B F , Dyke S J , Sain M K , and Carlson J D 1997 Phenomenological Model of a Magnetorheological Damper *Journal of Engineering Mechanic, ASCE,* **123(3),** 230–238.
[8] Yang S , Li S , Wang X , Gordaninejad F and Hitchcock G 2005 A Hysteresis Model for Magneto-rheological Damper *International Journal of Nonlinear Sciences and Numerical Simulation,* **6(2),** 139–144.
[9] Liu Y, Matsuhisa H, Utsuno H and Park J G 2006 Variable damping and stiffness control with magnetorheological dampers for a 2-DOF vibration system *International Journal of Japan Society of Mechanical Engineering (Series C),* **49 (1),** 156–162.

IOP Conf. Series: Materials Science and Engineering 1004 (2020) 012004 doi:10.1088/1757-899X/1004/1/012004

Role of thickness variation on the tailored bistability of unsymmetric composite laminates

P M Anilkumar[1], A Phanendra Kumar[1], A Haldar[2] and B N Rao[1*]
[1]Structural Engineering Division, Department of Civil Engineering, IIT Madras, Chennai, India
[2]School of Engineering, Cardiff University, Cardiff CF24 3AA, UK
[*]E-mail: bnrao@iitm.ac.in

Abstract. Unsymmetric composite laminates exhibit two stable configurations at room temperature after being cured at elevated temperature. Bistability of cross-ply laminates is due to the presence of residual strains imparted during the curing process. A simple snap-through process can be used to switch from one stable shape to another. The amount of snap-through force required for attaining the other stable shape depends on the total strain energy of the laminate in the first cured shape. Bistable shapes of the conventional square laminates are in the same energy level, which means the stable configurations are equally stable and results in equal snap-through and snap-back forces. By tailoring the energy levels of stable shapes, snap-through and snap-back forces can be varied. From various studies in the past, it is observed that bistable laminates have enormous potential for morphing applications. In some of the practical applications, one may require stable shapes at different energy levels, with unequal snap-through and snap-back energies for the shape transition. This study aims to tailor the energy levels of bistable shapes of the square laminate by varying the thickness of individual layers in the considered [0/90] square laminates. A detailed parametric study is performed in commercially available finite element software. It is concluded from the study that individual lamina thickness has a noticeable influence on the bistability of cross-ply laminates.

1. Introduction

Unsymmetrical composite laminates exhibit multiple equilibrium states when they are cooled from the curing temperature to the room temperature. The formation of multiple equilibrium states, rather than just a single equilibrium state, depends on the laminate dimensions, as well as the lamina fiber alignments and the material properties, specifically the mismatch of thermal expansion coefficients in the fiber direction between the layers. Hyer [1] in his pioneering works conducted a series of experiments [2] and formulated an analytical model [3, 4] to demonstrate the bistable behavior of thin unsymmetric composite laminates (two stable cylindrical equilibrium shapes and one unstable saddle shape) upon curing as shown in figure 1.

Since multiple equilibrium states are possible at the end of the curing process, the cured shape of unsymmetrical laminates can be changed from one stable state to another by appropriate snap-through mechanisms. Hyer modified his earlier model to precisely depict the behavior of thin cross-ply laminates considering geometric nonlinearities with Classical Laminate Theory (CLT) [5]. This anomalous behavior of unsymmetrical laminates attained much research interest, and many researchers contributed by improving Hyer's theory to take advantage of this unusual characteristic [6-11]. Thereafter, several analytical models have been reported in the literature to analyze the cured room-temperature shapes of unsymmetric laminates [12-19]. Analytical models developed by various researchers explained the difficulty predicting the exact cured shapes of the unsymmetric laminates, especially at its boundaries. To predict the shapes and snap through loads accurately, one requires a higher order of polynomial functions in the analytical model, which is computationally expensive.

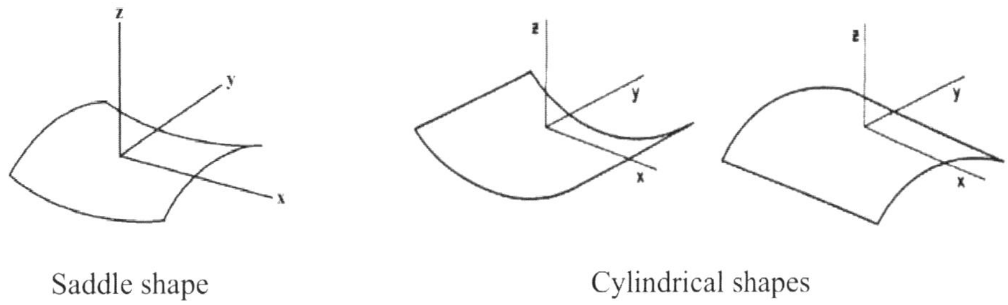

Figure 1: Equilibrium shapes of bistable laminates

Besides the analytical approaches, some of the researchers have also employed the finite element method to study the behavior of bistable plates. With refined FE models [20, 21], it is possible to predict the bistable shapes with higher accuracy, and they are closer to the experimentally observed shapes than the predictions from analytical models developed so far. Unlike analytical modelling, by using FE, one can model even the complex shapes using these bistable laminates. One of the main advantages of FE modelling is that it is possible to predict complex shapes generated by the bistable laminates numerically with at high accuracy [22-28].

The bistable nature of a structure can be explained based on its two distinct potential energy minima's. Each of the minima corresponds to one of the stable geometric shapes of the structure, and the graph for a bistable structure with equally stable shapes is shown in figure 2. From figure 2, one can observe that the structure has two stable states SS_1 and SS_2. External energy needs to be supplied to the structure to actuate between one stable shape and the other. When sufficient external energy is given in the form of actuation force, the potential energy of the structure increases until it reaches the curve's peak, and the structure will snap-through to another stable shape. If the external energy is not sufficient, snap-through will not occur, and the structure remains in the same stable shape. If the values SS_1 and SS_2 are equal, then the structure can have equilibrium shapes with equal stability, and if the values of SS_1 and SS_2 are not equal, then the structure can have equilibrium shapes with tailored stability. In such a case, the state with less energy is more stable compared to the other shape.

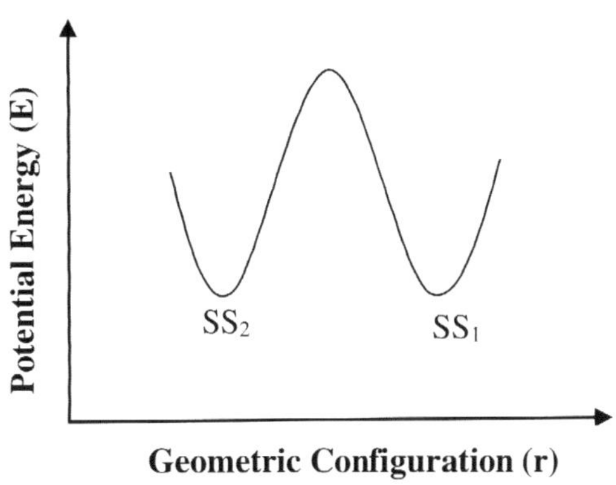

Figure 2: Potential energy curve (E: Potential Energy and r: Geometric Configuration)

In many practical applications like aerospace applications, there is a huge demand for structures that can undergo shape change based on the environmental conditions called adaptive structures. Bistable laminates have attained significant attention in the design of adaptive structures due to the low energy requirements for the shape transition. In some of the applications, the designer may have to use bistable shapes, where one stable shape is required to be more stable than the other. This helps in further reduction of energy consumption compared to the stable shapes, which are in equal energy levels (Figure 2). In 2002, Hufenbach et al. [29] incorporated multistable composites in adaptive structures by investigating various aspects like laminate edge length L, different angle-ply laminates, and variable layer ratio. The study showed the possibility of utilizing the residual stress field to generate monostable and multistable equilibrium states. Taking the studies by Hufenbach et al. [29] as a reference, a detailed parametric study is proposed in this study to understand the role of lamina thickness variations to tailor energy levels of bistable shapes of unsymmetric composite laminates. The proposed parametric investigation is performed in commercially available finite element software, Abaqus^{TM}. Details of the steps followed in the study are given in the subsequent sections.

2. Problem definition

Composite bistable shells obtained from the cool down process of [0/90] square laminate is used for the present analysis. The material properties of the laminate are given in table 1.

Table 1: Properties of laminate

Engineering constants		Value
Elastic Modulus	E_1	294 GPa
	E_2	9.5 GPa
Poisson Ratio	ν_{12}	0.3
Shear Modulus	G_{12}	5.0 GPa
Coefficient of thermal expansion	α_{11}	-2×10^{-8}/℃
	α_{22}	-2.25×10^{-5}/℃

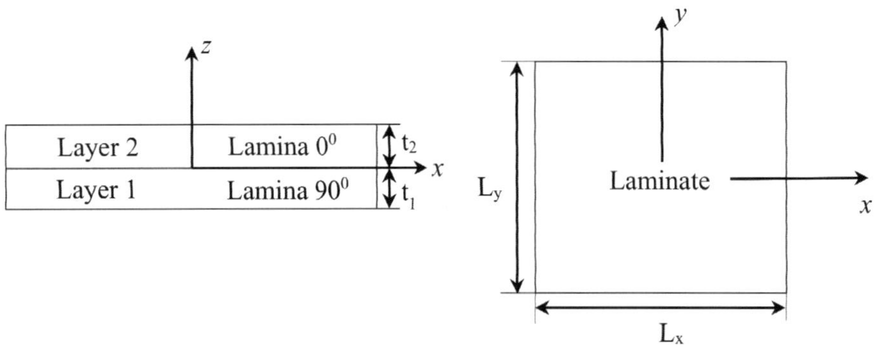

Figure 3: Laminate geometry

Figure 3 shows the schematic representation of the geometry used for the present analysis. Where t_1

is the thickness of lamina, which is fixed, t_2 is the thickness of lamina, which is varying for the study, and "p" is the ratio of the thickness (t_2/t_1). To investigate the effect of individual lamina thickness on the cured shape of laminate, one of the lamina thickness is increased from 0.100 mm to 0.150 mm while keeping the thickness of another lamina constant (0.100 mm). Five square geometries are taken for the numerical parametric study ($L=L_x=L_y$), L = 100 mm, 125 mm, 150 mm, 175 mm, 200 mm. Details of the proposed study are given in the following sections.

3. Numerical investigation

Thermally induced shapes of bistable unsymmetric composite laminates are modelled in a commercial finite element package, Abaqus™ for the analysis. Static non-linear analysis is performed by considering geometric non-linearities in the numerical model. Imperfections are introduced to the model of proposed laminate to avoid the convergence to unstable saddle shape during the analysis. Four-node quadrilateral general-purpose composite shell elements, type S4R in the Abaqus™ is taken for the finite element analysis. Mesh convergence studies are performed to find a suitable mesh characteristics for the described problem. Steps followed in the numerical investigation are explained in the subsequent subsections.

3.1 Cool-down step

The modelled plate is subjected to a curing temperature of 180° C during the initial stage. The temperature field of the laminate assembly is then brought down to room temperature at 20° C. The assigned temperature difference produces residual stresses, which makes the modelled plate deviate to any one of its stable shapes.

3.2 Snap-through and stability check

The static transverse load is applied on the corners of the modelled cured shape of laminate. Load higher than the required snap-through are applied so that the laminate will deform to another equilibrium shape. An additional step (stability check) is imposed to ensure the stability of obtained stable state even after the removal of the applied load.

4. Results and discussion

Figure 4 shows the cured bistable shapes of an unsymmetric cross-ply laminate. Figure 5 shows the variation of out of plane displacement with curing temperature for lamina with 150 mm side length. A noticeable increase in maximum out-of-plane displacement of the other stable shape, even after the removal of snap-through load, is observed with the increase in thickness ratio. Strain energy levels of both stable shapes are extracted from the developed numerical model.

Figure 4: Modelled bistable shapes

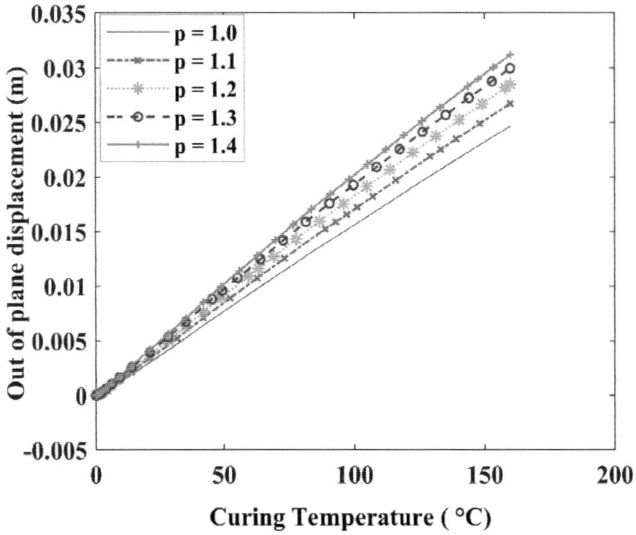

Figure 5: Out of plane displacement with curing temperature for L=150 mm

Table 2: Difference in SE with thickness variation

p	Difference in SE				
	L=100 mm	L=125 mm	L=150mm	L=175mm	L=200mm
1.0	0.00	0.00	0.00	0.00	0.00
1.1	4.85	5.59	5.15	5.36	5.30
1.2	9.52	9.76	10.59	10.56	10.45
1.3	14.15	15.06	15.00	15.34	15.46
1.4	18.69	19.05	19.83	20.36	20.47
1.5	-	-	-	-	25.12

Table 3: Ratio of snap forces with thickness variation

p	Ratio of snap force				
	L=100 mm	L=125 mm	L=150mm	L=175mm	L=200mm
1.0	1.01	1.01	1.01	1.01	1.01
1.1	1.43	1.33	1.37	1.40	1.45
1.2	2.45	2.30	2.47	2.16	2.20
1.3	5.20	4.53	4.98	4.82	4.07
1.4	25.26	15.19	13.82	10.28	9.31
1.5	-	-	-	-	12.50

Table 2 and Table 3 show the difference in strain energy (SE) and the ratio of snap forces (which is defined as the ratio between snap-through and snap-back forces), respectively with the thickness variation. There is a noticeable difference in SE with the increase of thickness of one lamina. It shows the presence of tailoring in bistability. The two stable shapes obtained as a result of the cool-down process seem to be in different energy levels. The above observation shows a higher degree of stability for one of the stable shapes. Calculated snap-through and snap-back forces are given in table 3.

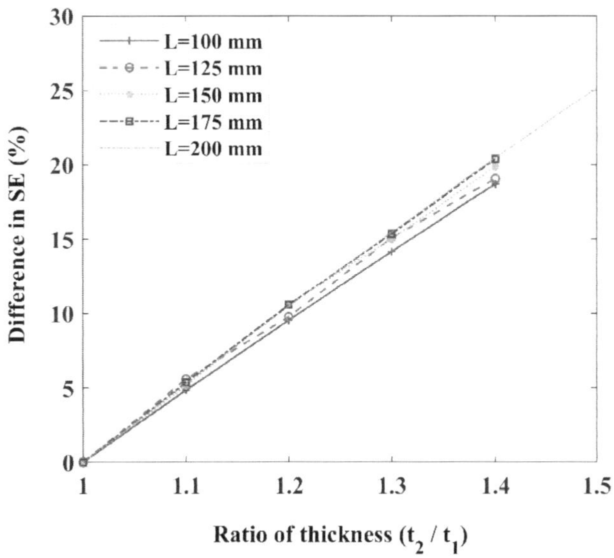

Figure 6: Difference in SE with thickness variation

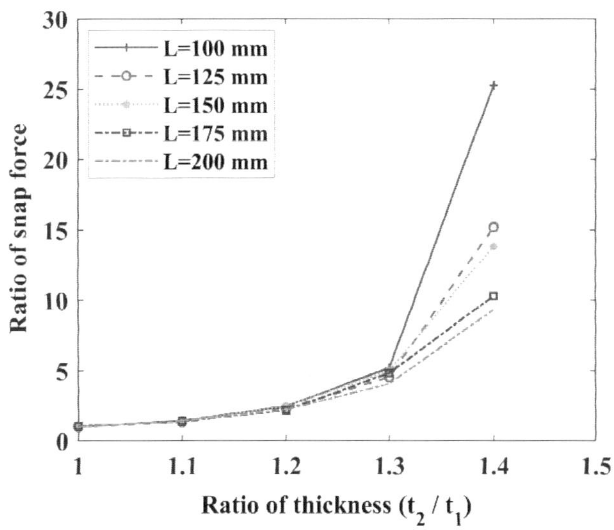

Figure 7: Ratio of snap forces with thickness variation

There is a noticeable difference in SE with the increase of thickness of one lamina. It shows the presence of tailoring in bistability. The two stable shapes obtained as a result of the cool-down process seem to be in different energy levels. The above observation shows a higher degree of stability for one of the stable shapes. The effect of change in "p" on the ratio between the snap-through/snap-back is shown in table 3. Bistability of the laminate is lost when "p" is 1.5 for L = 100 mm, 125 mm, 150 mm and 175 mm. Graphical representation of table 2 and table 3 are given in figure 6 and figure 7, respectively. After a specific point (t_2 = 0.150 mm), loss in bistability is observed. Figure 8 shows the schematic representation of energy levels for equally stable shapes and tailored stable shapes of the square laminate.

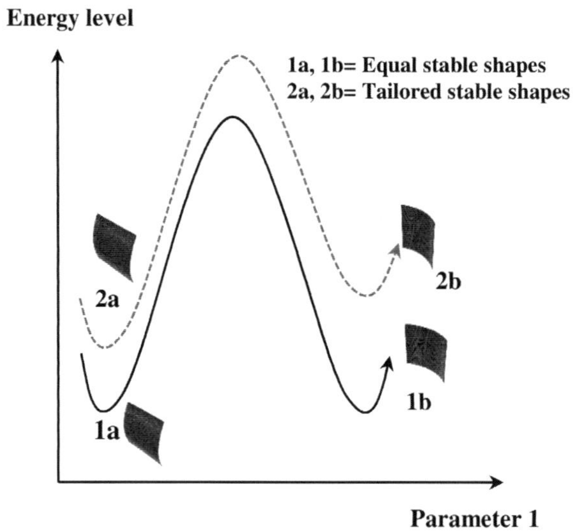

Figure 8: Schematic representation of energy levels

5. Conclusion

With the increase in the thickness of individual lamina, the stiffness of the laminate gets increases. It is observed that after a specific point (limit of t_2/t_1 = 1.5), laminate loses its bistability completely. From the ratio of snap forces, it is observed that the laminate with L = 100 mm is having a more stable second shape compared to the first shape. This is due to the higher snap-back force compared to the snap-through force in L = 100 mm compared to other different length cases considered. It can be concluded that in order to have laminate with tailored bistability, a bistable laminate with shorter length is more preferred. The focus needs to be laid on the thickness ratio of the laminates being used for this application as we may lose the bistability completely.

Acknowledgement

The first author would like acknowledge Prime Ministers Research Fellowship Scheme (PMRF), India for the support during the course of this research.

References

[1] Hyer M W 1981 Some Observations on the Cured Shape of Thin Unsymmetric Laminates *Journal of Composite Materials*, **15**, 175-194

[2] Hyer M W 1981 Calculation of the Room- Temperature Shapes of Unsymmetric Laminates *Journal of Composite Materials*, **15**, 296-310.

[3] Hyer M W 1981 The Room-Temperature Shapes of Four-Layer Unsymmetric Cross-Ply Laminates *Journal of Composite Materials*, **16**, 318-340.

[4] Hyer M W and Bhavani P C 1984 Suppression of Anticlastic Curvature in Isotropic and Composite Plate *International Journal of Solids and Structures*, **20**, 553–570.

[5]Hamamoto A and Hyer M W 1987 Nonlinear temperature-curvature relationships for unsymmetric graphite-epoxy laminates *International Journal of Solids and Structures*, **23**, 919-935.
[6]Dang J and Tang Y 1986 Calculation of the room- temperature shapes of unsymmetric laminates *Proceedings of the international symposium on composite materials and structures ed T T Loo and C T Sun* (Technomic Publishing Co Inc) 201-207.
[7]Jun W J and Hong C S 1992 Cured shape of unsymmetric laminates with arbitrary lay-up angles *Journal of Reinforced Plastic and Composites*, **11**, 1352–66.
[8]Peeters L J B, Powell P C and Warnet L 1996 Thermally-induced shapes of unsymmetric laminates *Journal of Composite Materials*, **30**, 603–26.
[9]Dano M L and Hyer M W 1996 Room-temperature shape of general unsymmetric laminates *Proceedings of the 11th Technical Conference of the American Society for Composites* (Virginia Polytechnic Institute and State University, USA, October 7–9, 1996) 100-110.
[10] Etches J, Potter K, Weaver P and Bond I 2009 Environmental effects on thermally induced multistability in unsymmetric composite laminates *Composites Part A: Applied Science and Manufacturing*, **40**, 1240-1247.
[11] Dano M L and Hyer M W 1998 Thermally-induced deformation behavior of unsymmetric laminates *International Journal of Solids and Structures*, **35**, 2101-2120
[12] Pirrera A, Avitabile D and Weaver P M 2010 Bistable plates for morphing structures: a refined analytical approach with high-order polynomials *International Journal of Solids and Structures*, **47**, 3412-3425.
[13] Dano M L and Hyer M W 1996 The response of unsymmetric laminates to simple applied forces *Mechanics of Composite Materials and Structures*, **03**, 65-80.
[14] Hufenbach W and Gude M 2002 Analysis and optimisation of multistable composites under residual stresses *Journal of Composite Structures*, **55**, 319-327.
[15] Hufenbach W, Gude M and Kroll L 2002 Design of multistable composites for application in adaptive structures *Journal of Composites Science and Technology*, **62**, 2201-2207.
[16] Hufenbach W, Gude M, Kroll L, Sokolowski A and Werdermann B 2001 Adjustment of residual stresses in unsymmetric fiber reinforced composites using genetic algorithms *Mechanics of Composite Materials,* **37**, 71-72.
[17] Hyer M W and Jilani A 1998 Predicting the deformation characteristics of rectangular unsymmetrically laminated piezoelectric materials *Smart Materials and Structures*, **07**, 784-791.
[18] Betts D N, Salo A I T, Bowen C R and Kim H A 2010 Characterisation and modelling of the cured shapes of arbitrary layup bistable composite laminates *Journal of Composite Structures*, **92**, 1694-1697.
[19] Haldar A, Groh R M J, Jansen E and Weaver P M 2020 An efficient semi-analytical framework to tailor snap-through loads in bistable variable stiffness laminates International *Journal of Solids Structures*, **195**, 91-107.
[20] Schlecht M and Schulte K 1999 Advanced calculation of the room-temperature shapes of unsymmetric laminates *Journal of Composite Materials*, **33**, 1472-1490.
[21] Schlecht M, Schulte K and Hyer H W 1995 Advanced calculation of the room-temperature shapes of thin unsymmetric composite laminates *Journal of Composite Structures*, **33**, 627-633.
[22] Haldar A, Reinoso J, Jansen E and Rolfes R 2018 Thermally induced multistable configurations of variable stiffness composite plates: Semi- analytical and finite element investigation *Journal of Composite Structures*, **183**, 161-175.
[23] Anilkumar P M, Haldar A, Jansen E, Rao B N and Rolfes R 2019 Design Optimization of Multistable Variable-Stiffness Laminates *Mechanics of Advanced Materials and Structures*, **26**, 48–55.

[24] Anilkumar P M, Haldar A, Jansen E, Rao B N and Rolfes R 2018 Effect of actuation procedure in MFC actuators for morphing of bistable laminates *Proceedings of 11th Structural Engineering Convention* (Jadavpur University, Kolkata, India) 37-42.
[25] Haldar A, Jansen E and Rolfes R 2019 Analysis of Morphing Trailing Edge Flap with Embedded Multistable Variable Stiffness Laminates *AIAA Scitech Forum* pp 0856.
[26] Anilkumar P M, Rao B N, Haldar A, Jansen E and Rolfes R 2019 Effect of Piezoelectric Fibre Alignment on Morphing of Bistable Composites Proceedings of 29th International Workshop on Computational Mechanics of Materials (University of Zagreb, Dubrovnik, 15-18 September 2019).
[27] Attada P K, Parapparambil Muraleedharan Nair A and Bhairavavajjula N R 2020 Study on the actuation force of triangular bistable composite laminates Preprint https://doi.org/10.1002/mdp2.169
[28] Haldar A, Jansen E, Hofmeister B, Bruns M and Rolfes R 2020 Analysis of Novel Morphing Trailing Edge Flap Actuated by Multistable Laminates *AIAA Journal*, **58**, 3149-3158.
[29] Hufenbach W, Gude M and Kroll L 2002 Design of multistable composites for application in adaptive structures Composites Science and Technology, **62**, 2201-2207.

IOP Conf. Series: Materials Science and Engineering 1004 (2020) 012005 doi:10.1088/1757-899X/1004/1/012005

A Comparative Study on Vibration-Based Damage Identification and Localization Methods

A J Shah[1*] V Jayprakash[2] and G R Vesmawala[3]

[1]Research Scholar, Civil Engineering Department, S.V. National Institute of Technology, Surat – 395007, Gujarat, India.

[2]M.Tech Student, Civil Engineering Department, S.V. National Institute of Technology, Surat – 395007, Gujarat, India.

[3]Associate Professor, Civil Engineering Department, S.V. National Institute of Technology, Surat-395007, Gujarat, India

[*]E-mail: ankurphd1718@gmail.com

Abstract. This paper outlines various vibration-based methods for detecting and locating damage in mechanical and civil structural systems. The basic premises of these methods are that any changes in physical characteristics (mass and stiffness) of the structure will induce detectable changes in the modal properties of the structure (natural frequency, mode shape and damping). A comparative study of eight extensively used damage identification methods is done on a simulated simply supported beam to demonstrate the validity and effectiveness. The outcomes of various methods are compared in terms of damage detection and localization under single and multiple damage scenarios.

1. Introduction

The traditional damage identification methods like visual inspection and non-destructive testing are less efficient since they require the location of damage to be easily known and readily accessible for testing purposes. Vibration-based techniques for damage identification were developed due to the need for identification of damage globally. The fundamental premise for such global identification method is that vibration characteristics such as natural frequency, mode shape and damping of the structure will be affected by the change in physical properties of the structure such as mass and stiffness. Hence, it is intuitive to examine the changes in structure's modal properties to detect the damage [1].

The method for identifying damage can be categorized in to following four levels [2] based on the damage information provided by the methods

1) Level 1: Damage Existence Identification
2) Level 2: Damage Location Identification
3) Level 3: Damage Severity Identification
4) Level 4: Remaining Life Prediction

There is no unique vibration-based damage identification method which gives accurate results for all structural systems and levels of damage. Even though different methods yield varying results, it is expected that these results converge thereby increasing the confidence in output. In this paper, the aspects under consideration for various damage identification methods are damage existence (Level 1) and localization (Level 2). The selected methods for comparison are: MAC & COMAC methods [3];

Mode Shape Difference method [4]; Mode Shape Curvature method [5]; Cumulative Damage Factor method [6]; Damage Index method [7]; Modal Flexibility Change method [8]; and Uniform Flexibility Shape Curvature Change method [9].

2. Vibration-Based Damage Identification Methods

2.1 Modal Assurance Criterion (MAC) and Coordinate MAC (COMAC) Method

MAC is a scalar quantity having values ranging 0 to 1 which indicates the resemblance between mode shapes while COMAC values indicates a point wise change between the mode shapes and it ranges between 1 and 0.

$$MAC = \frac{\sum_{i=1}^{m} | \phi_i^T \phi_i^* |^2}{\sum_{i=1}^{m} | \phi_i \phi_i^T | \sum_{i=1}^{m} | \phi_i^* \phi_i^{*T} |} \tag{1}$$

$$COMAC_j = \frac{| \sum_{i=1}^{m} \phi_{i,j} \phi_{i,j}^* |^2}{\sum_{i=1}^{m} (\phi_{i,j})^2 \sum_{i=1}^{m} (\phi_{i,j}^*)^2} \tag{2}$$

where ϕ = mode vectors of intact beam; ϕ^* = mode vectors of damaged beam; m = number of modes considered; i = mode number; and j = node number.

2.2 Mode Shape Difference Method

This traditional method makes use of the normalized absolute differences of the modal vectors of intact and damaged structures

$$\Delta\phi_{i,j} = | \phi_{i,j}^* - \phi_{i,j} | \tag{3}$$

$$MSD_j = \sum_{i=1}^{m} \Delta\phi_{i,j} \tag{4}$$

2.3 Mode Shape Curvature (MSC) Method

MSC is the second derivative of mode shape and it is highly sensitive to the damage in the structure. Applying the Central Difference approximation over the displacement mode shapes, Curvature at node j along the beam is given by

$$\phi_j'' = \frac{\phi_{j+1} - 2\phi_j + \phi_{j-1}}{h^2} \tag{5}$$

Where ϕ_j''= Curvature at node j; and h = length of each element.

The absolute curvature difference between the intact and damage structure is given as

$$\Delta MSC_j = | \phi^{*''}_j - \phi_j'' | \tag{6}$$

2.4 Cumulative Damage Factor (CDF) Method

CDF is the average of absolute difference between the intact and damaged curvature mode shapes. CDF of jth node, considering 'm' number of modes is given by:

$$CDF_j = \frac{1}{m} \sum_{i=1}^{m} \Delta MSC_j \tag{7}$$

2.5 Damage Index Method (DI) Method
The damage index method depends on the change of strain Energy which is stored in the structure when it deforms under a particular mode shape. For a location j on the beam the change in the ith mode, strain energy is related to changes in curvature of the mode at location j. The damage index for this location and this mode β_{ij}, is defined as

$$\beta_i = \frac{\left[\int_a^b [\phi_i^{*''}(x)]^2 dx + \int_0^L [\phi_i^{*''}(x)]^2 dx\right] \int_0^L [\phi_i^{''}(x)]^2 dx}{\left[\int_a^b [\phi_i^{''}(x)]^2 dx + \int_0^L [\phi_i^{''}(x)]^2 dx\right] \int_0^L [\phi_i^{*''}(x)]^2 dx} \quad (8)$$

The normalized damage index Z is given by

$$Z_i = \frac{\beta_i - \mu}{\sigma} \quad (9)$$

where μ = mean of β_i; and σ = standard deviation of β_i. To reduce the problem of false alarms due to influence of higher modes, normalized mode shape curvatures are used [10].

2.6 Modal Flexibility Change Method
The flexibility matrix [F] is given by

$$F = [\phi] * [\frac{1}{\omega^2}] * [\phi^T] \quad (10)$$

Where $[\phi]$ = mass orthonormalized mode shape matrix; and ω = natural frequency.

The change in Flexibility matrix for a structure can be given by

$$[\Delta F] = [F] - [F^*] \quad (11)$$

Where [F]= flexibility matrix of intact beam; and $[F^*]$ =flexibility matrix of damaged beam.

The Element with the largest absolute value in the kth column of ΔF is considered as the kth element of the row vector $[\delta_{max}]$.

$$\delta_{\max k} = \max |\Delta F|_k \quad (12)$$

2.7 Uniform Flexibility Shape Curvature Change Method
The change in flexibility curvature is given by the equation

$$\Delta F'' = [F''] - [F^{*''}] \quad (13)$$

Where $[F'']$= Curvature of flexibility matrix of intact beam; and $[F^{*''}]$ = Curvature of flexibility matrix of damaged beam.

The Element with the largest absolute value in the kth column of $\Delta F''$ is considered as the kth element of the row vector $[\delta''_{\max k}]$.

$$\delta''_{\max k} = \max|\Delta F''|_k \quad (14)$$

3. Methodology

A simply supported beam of rectangular cross-section (0.25m (width) x 0.2m(depth)) having 4m length was selected for this study. The modulus of elasticity and density of the beam were presumed to be 32 GPa and 2500 kg/m^3 respectively. MATLAB programming was used to obtain the modal properties by considering beam as 2-D Euler-Bernoulli beam. The beam was divided equally into 20 2-D beam elements (Figure 1). All the mode shapes obtained were normalized to the mass matrix. Different damage scenarios were added by reducing the stiffness i.e. modulus of elasticity of the beam elements as shown in Table 1. To obtain the assumed experimental modal parameters, modal analysis was again done in MATLAB using the properties of damaged beam. The data obtained was further used in various damage detection algorithms to detect and locate the damage.

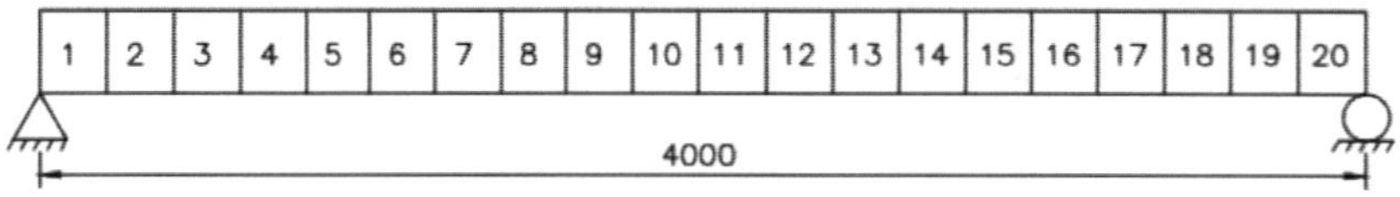

Figure 1: Simulated Simply Supported Beam

Table 1: Damage Scenarios

Damage Cases	D0	D1	D2	D3	D4	D5	D6
Damaged element	-	3	10	15	10,11	3,18	3,10,11,18
Severity	-	10	50	90	20%, 30%	40%, 30%	20%, 40%, 30%, 30%

4. Results and Discussions

The natural frequencies of intact and various damaged beams are given in Table 2. The reduction in natural frequencies indicates the existence of the damage but the exact location of damage is not known. COMAC values in each mode (Table 3) and MAC values of each nodes (Table 4) also serves the purpose of damage detection but gives no indication of location of damage.

Table 2: Natural Frequencies of beam under different damage scenarios

Modes	D0	D1	D2	D3	D4	D5	D6
1	20.28	20.23	20.06	20.17	19.86	18.56	19.32
2	81.13	80.55	81.02	80.27	76.34	80.26	74.73
3	182.61	180.82	181.13	181.63	169.55	171.91	175.91
4	325.00	322.24	323.67	324.56	307.38	315.46	321.04
5	508.98	506.31	506.26	506.23	492.01	490.10	488.93

Table 3: COMAC values in different damage scenarios

Nodes	D1	D2	D3	D4	D5	D6
1	0.000	0.000	0.000	0.000	0.000	0.000
2	1.000	0.997	0.995	1.000	0.999	1.000
3	1.000	0.997	0.993	1.000	1.000	1.000
4	1.000	0.998	0.986	0.999	1.000	0.999
5	1.000	0.998	0.982	0.999	0.999	1.000
6	1.000	0.996	0.978	0.999	0.999	0.999
7	1.000	0.997	0.966	0.998	1.000	0.998
8	1.000	0.997	0.966	0.999	1.000	0.998
9	1.000	0.995	0.943	0.999	1.000	0.996
10	1.000	1.000	0.911	0.999	1.000	0.998
11	1.000	1.000	0.927	1.000	0.999	1.000
12	1.000	0.994	0.930	0.999	0.999	0.998
13	1.000	0.997	0.912	0.998	0.998	0.999
14	1.000	0.997	0.913	0.999	0.998	1.000
15	1.000	0.994	0.928	0.999	0.997	0.998
16	1.000	0.997	0.989	0.999	0.997	0.999
17	1.000	0.997	0.980	0.999	0.996	0.999
18	1.000	0.998	0.970	1.000	0.998	1.000
19	1.000	0.998	0.964	1.000	0.999	1.000
20	1.000	0.996	0.965	1.000	0.998	0.999

21	0.000	0.000	0.000	0.000	0.000	0.000

Table 4: MAC values of different damage scenarios in first five modes

Modes	D1	D2	D3	D4	D5	D6
1	1	1	0.999	1	1	1
2	1	1	0.999	1	0.999	1
3	0.999	0.999	0.998	0.999	0.999	0.999
4	0.999	0.999	0.996	0.999	0.999	0.999
5	0.999	0.999	0.988	0.999	0.999	0.999

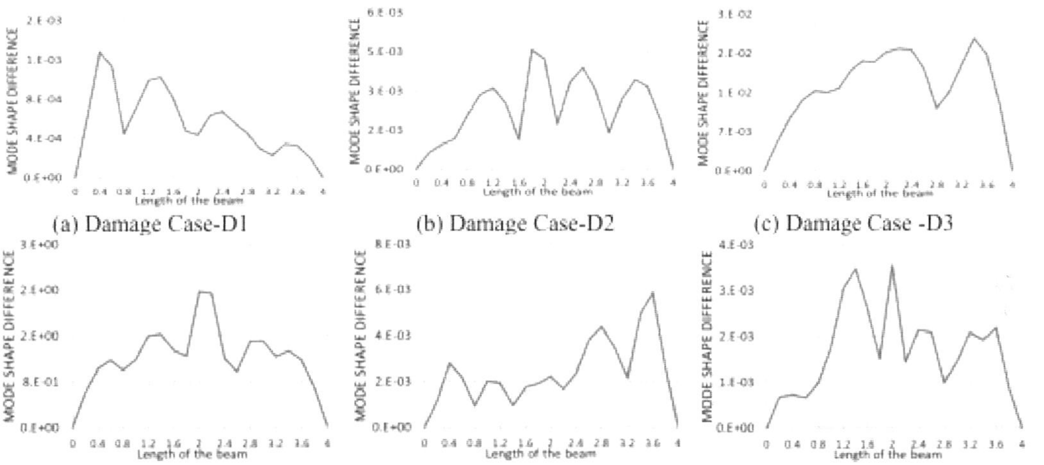

(a) Damage Case-D1 (b) Damage Case-D2 (c) Damage Case -D3
(d) Damage Case-D4 (e) Damage Case -D5 (f) Damage Case -D6

Figure 2: (a)-(f) MSD in damage scenarios D1-D6 considering first three modes

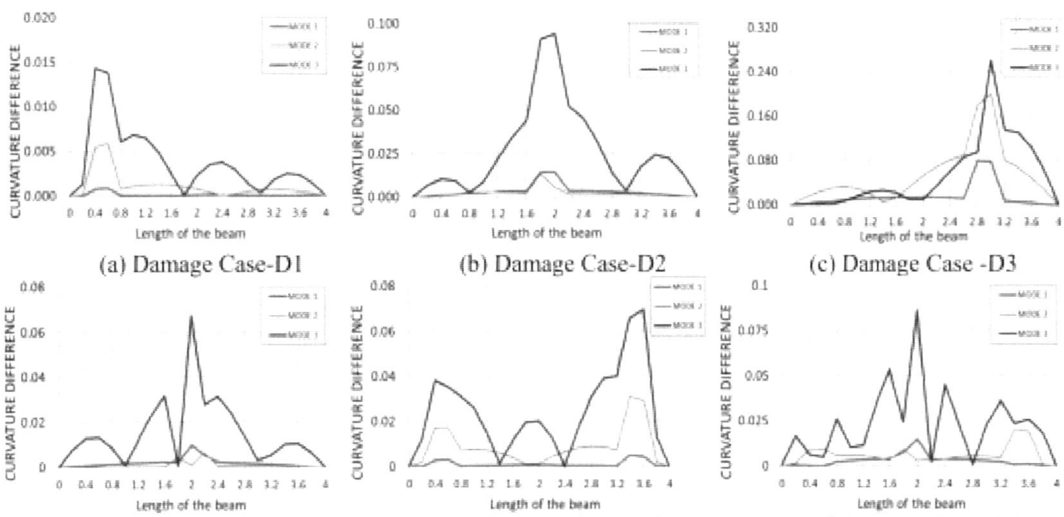

(a) Damage Case-D1 (b) Damage Case-D2 (c) Damage Case -D3
(d) Damage Case-D4 (e) Damage Case -D5 (f) Damage Case -D6

Figure 3: (a)-(f) MSC difference in first three modes for D1-D6 damage scenarios.

IOP Conf. Series: Materials Science and Engineering 1004 (2020) 012005 doi:10.1088/1757-899X/1004/1/012005

The absolute differences in Mode shapes (MSD) for various damage scenarios are plotted in Figure 2. It is indicated that absolute difference of mode shape is more in damaged region. This method gives many false indications in multiple damage scenarios.

The absolute differences in Mode Shape Curvatures (MSC), between intact and various damaged beams for Mode-1, Mode-2 and Mode-3 are plotted in Figures 3. For damage cases D1-D5, it was observed that at lower modes (mode 1 and mode 2) damage locations are indicated by definite peaks in graphs, but at higher modes, numerous peaks are observed leading to false positive alarms. Cumulative Damage Factors (CDF) of damaged and intact beams considering Mode-1, Mode-2 and Mode-3 are plotted in Figure 4. Both MSC and CDF method successfully predict single damage location. For Multiple damage case D6, both MSC and CDF methods do not exhibit good performance.

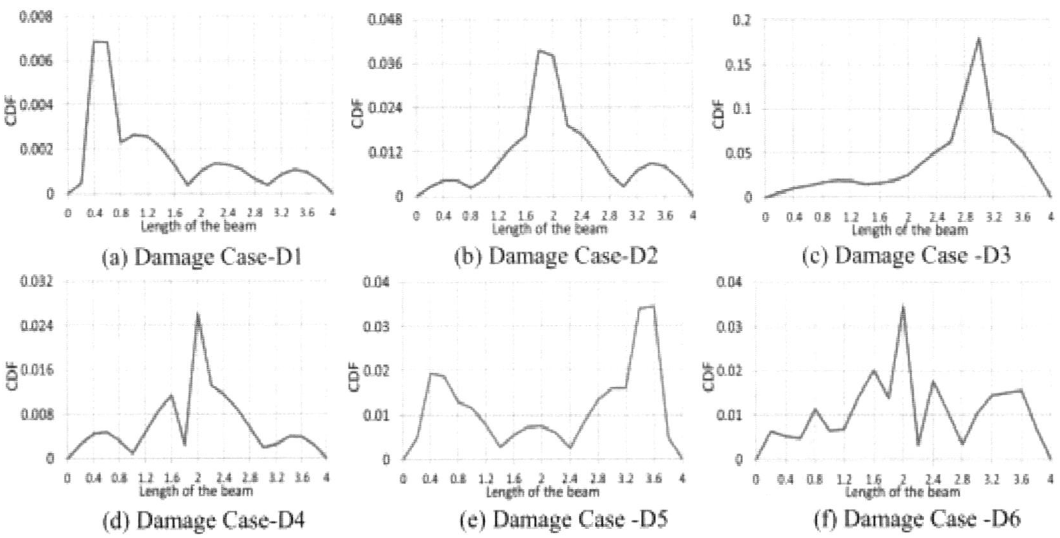

Figure 4: (a)-(f) CDF D1-D6 damage scenarios considering first three modes

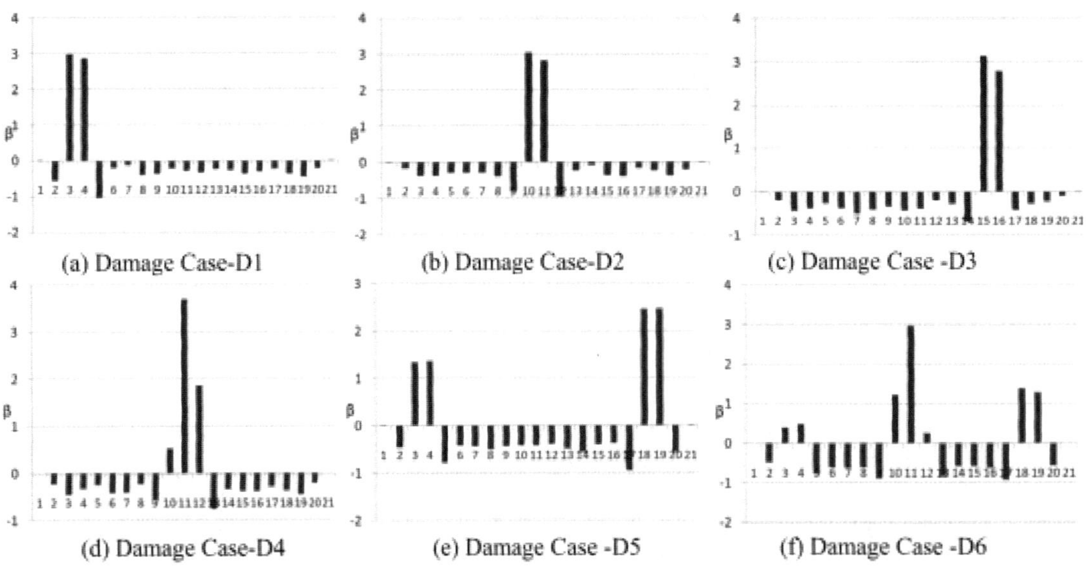

Figure 5: (a)-(f) Damage Index β for D1- D6 damage scenarios

The normalized damage index is shown in Figure 5. The positive values of index β indicate the presence of damage. The results indicate that the damage index method successfully predicts damage location for single and multiple damage scenarios.

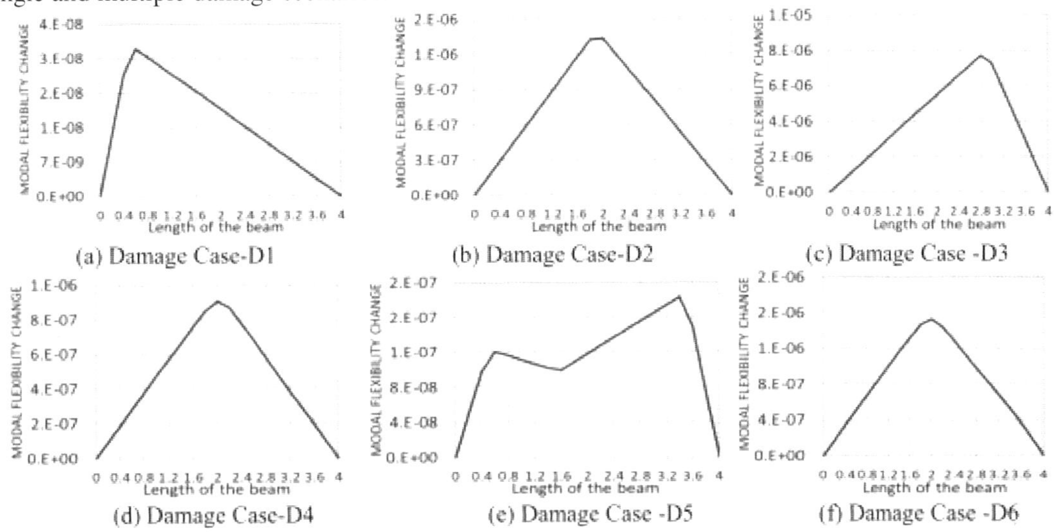

(a) Damage Case-D1 (b) Damage Case-D2 (c) Damage Case -D3

(d) Damage Case-D4 (e) Damage Case -D5 (f) Damage Case -D6

Figure 6: (a)-(f) Change in Modal Flexibility Matrix for D1- D6 damage scenarios

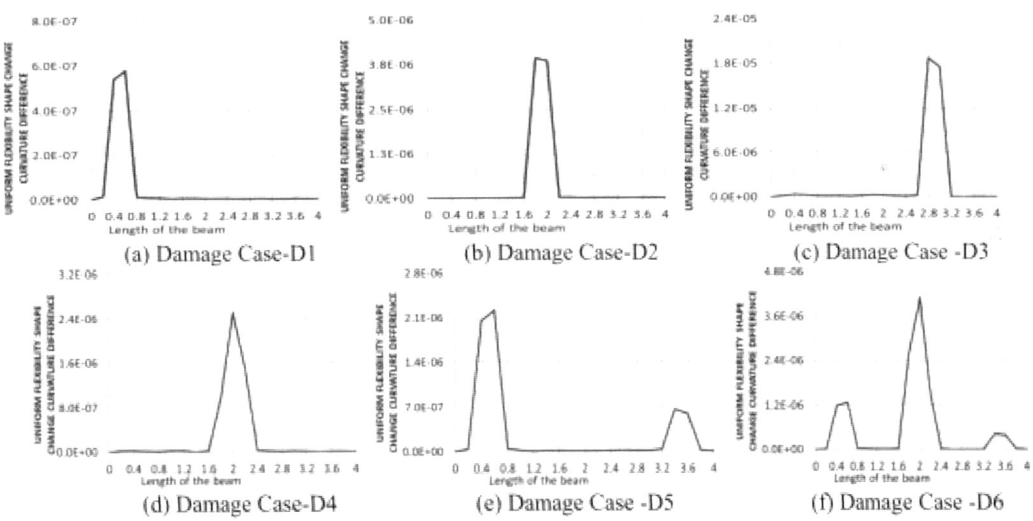

(a) Damage Case-D1 (b) Damage Case-D2 (c) Damage Case -D3

(d) Damage Case-D4 (e) Damage Case -D5 (f) Damage Case -D6

Figure 7: (a)-(f) Changes in Uniform Flexibility Shape Curvatures for D1- D6

The changes in flexibility between intact and different damaged beams are plotted in Figure 6. Damage locations are indicated accurately by peaks. It gave good results in double damage cases indicating damages by change in slopes. In case of multiple damages, changes in slope are not noticeable. The changes in uniform flexibility shape curvature are plotted in Figure 7. The peaks accurately indicate the location of damage for single and multiple damage scenarios.

5. Conclusions

A comparative study on simulated simply supported beam was carried out to evaluate the effectiveness of different algorithms for damage detection. The following conclusion are drawn from comparison among different damage identification methods.

1. The change in natural frequencies, MAC, COMAC and Mode shape difference methods detected the presence of damage but were unsuccessful in identifying the location of damage and hence can be classified as Level-1 methods.
2. Mode shape curvature and Curvature damage factor method could locate damages at lower modes but the results were ambiguous for higher modes. Also, both methods proved effectively only for single damage detection. For double and multiple damage scenarios both the methods do not exhibit good performance.
3. The change in Flexibility matrix method detected the location of damage for single and double damage scenarios. However, the results for multiple damage scenarios were not easily discernible.
4. Both Damage Index and uniform flexibility shape curvature change method proved to be effective in identification of damage in single and multiple damage scenarios.

References

[1] Doebling S W, Farrar C R, and Prime M B 1998 *Shock and Vibration digest.,* **30** 91-105
[2] Rytter A 1993 *Doctoral Dissertation* (Aalborg University).
[3] Allemang R J and Brown D L 1982 *Proc. Int. Modal Analysis Conf,* 110
[4] Ho Y K and Ewins D J 2000 *Proc. Int. conf. on system identification and structural health monitoring*, 677.
[5] Pandey A K, Biswas M and Samman, M M 1991 *Journal of Sound and Vibration,* **145** 321-332
[6] Wahab M A and De Roeck, G 1999 *Journal of Sound and Vibration*, **226** 217-235
[7] Stubbs N, Kim J T, and Farrar C R 1995 *Proc. 13th Int. Modal Analysis Conf,* 210
[8] Pandey A K and Biswas M 1994 *Journal of Sound and Vibration,* **169** 3-17
[9] Catbas F N, Gul M, and Burket J L 2008 *Smart material and Structures,* **17** 1-12
[10] Dawari V B, Kamble P P and Vesmawala G R 2015 *Advances in Structural Engineering,* **3** 2599-2607

IOP Conf. Series: Materials Science and Engineering 1004 (2020) 012006 doi:10.1088/1757-899X/1004/1/012006

CFD Investigation of Impact of Vessel configuration and Different Impeller Types in Stirred Tank

Mohit Pathak[1], **Thiyam Tamphasna Devi**[2*]

[1,2]Department of Civil Engineering, National Institute of Technology, Langol – 795004, Manipur
*Email: thiyam85@gmail.com

Abstract. Computational Fluid Dynamics (CFD) study is carried out to simulate the flow characteristics of different impeller types in stirred tanks of baffled and un-baffled system. The hydrodynamic behavior induced by a Rushton turbine (RT6) and Curve Blade impeller (CD-6) is numerically analysed by solving the Navier-Stokes equations coupling with k-ε turbulence model with Multi-reference frame (MRF) impeller model. The predicted results were validated with available experimental published data and show a satisfactory agreement. The power characteristics for the two agitator systems (baffled and un-baffled) were obtained for the turbulent system of the fluid flow. In these conditions, it is found that the power number depended strongly on the length of the baffle when baffled is provided. Therefore, the presence of the baffles in the tank greatly improved the mixing performance of the system but consumes higher power than the un-baffled system. Similarly, mixing improves when Rushton impeller is employed but consumes higher power than the Curve Blade impeller. Hence, it was observed that overall efficiency of the system is better when baffled system of Curve Blade impeller is opted in any mixing operation.

1. Introduction

The capability of Computational fluid dynamics (CFD) as a powerful tool for the design of and scale-up of mechanically stirred reactors has been reported by many researchers e.g. [1-4]. In recent years, CFD techniques are being increasingly used as a substitute for experiments to obtain the detailed flow field for a given set of fluid, impeller and tank geometries [5]. Computational fluid dynamics (CFD) is an accepted and well-used means of assessing and optimizing process designs without necessarily incurring the expense of prototype development [6]. Advantage of CFD based prediction methods is that it does not have scaling up or scaling down problems as it solves the fundamental equations governing fluid flow [4]. CFD is a powerful tool for investigating flows at low cost when it is required by a high-quality experimental facility [7].It is suggested by [2] that for accurately predicting the turbulence in the system, 'computationally intensive' conditions must be adopted but for studying the mean fluid flow 'computationally light' conditions may be applied in the simulation.

Most industrial stirred vessels are often used baffles in the system so that breaking of the undesired tangential velocity components can occur avoiding the pronounced central vortex formation at high angular speeds. It is expressed by [8] that the presence of baffles increases the circulation of velocities (axial and radial) which enhances better mixing performance in the system than the unbaffled system. Even though baffled systems are commonly used in several applications, for specific purposes unbaffled systems are also desirable. It is argued that if baffled is removed, the whole fluid flow characteristics will change and hence impacting to the performance of the system [9]. Therefore, Unbaffled systems are used for specific purposes if the presence of baffled results into drawbacks [10]. It is found out by [8] one drawback of baffled system that it actually worsens the mixing performance in the aeration process because of formation of dead zones at certain regions in the system. For example, in the food and pharmaceutical industries, it is necessary to keep the system clean as much as possible therefore formation of dead zone is undesirable and in biological applications [11] shear and impact are to be reduced upto certain limit. Likewise, in crystallization processes [12] and; in precipitation processes baffles may suffer from incrustation problems [13]. Scientific literature shows an increasing interest towards unbaffled stirred vessels: both experimental investigations [14-20] and computational studies [21-23, 9,10, 19]. It was found by [19] that the minimum power requirements for complete suspension in top-covered unbaffled vessels are lower than those in corresponding baffled tanks.

Standard impeller (Rushton) is commonly used but several researchers have shown its inferiority in performance [24-33] as compared with other types of impeller. Concave type blade was introduced by [35] modifying from the flat type blade keeping in mind to improve the performance of the system. Consequently, less than about 50% of the installed power of the motor is put to effective use during normal operations in such curve/concave blade type.

The objective of this work is to employ CFD technique to study the mixing operation of different impeller designs. It was also anticipated that this work would give an essential basis for accurate scale-up of mixing systems in industrial, field, and various type of agricultural mixing operations. In this study, different types of impellers: Curve blade and Rushton turbine impellers were designed and studied to deduce how their different design attributes will impact the flow characteristics. Velocity profiles generated in CFD were interpreted as the impeller flow characteristics. The standard k-epsilon (ε) model was used for modelling the turbulent flow process and the multiple reference frames (MRF) technique was employed to model the impeller motion in a baffled tank.

2. Stirred tank configuration

The first configuration we considered is that of a cylindrical tank equipped with a Rushton turbine (Fig.1a). The second is a fully baffled tank (Fig.1b). The height of the cylindrical tank is equal to its diameter (T=H), and the baffle thickness is defined by w=H/10. Impeller is placed in the axial position corresponding to z=H/3, and has diameter d=T/3.

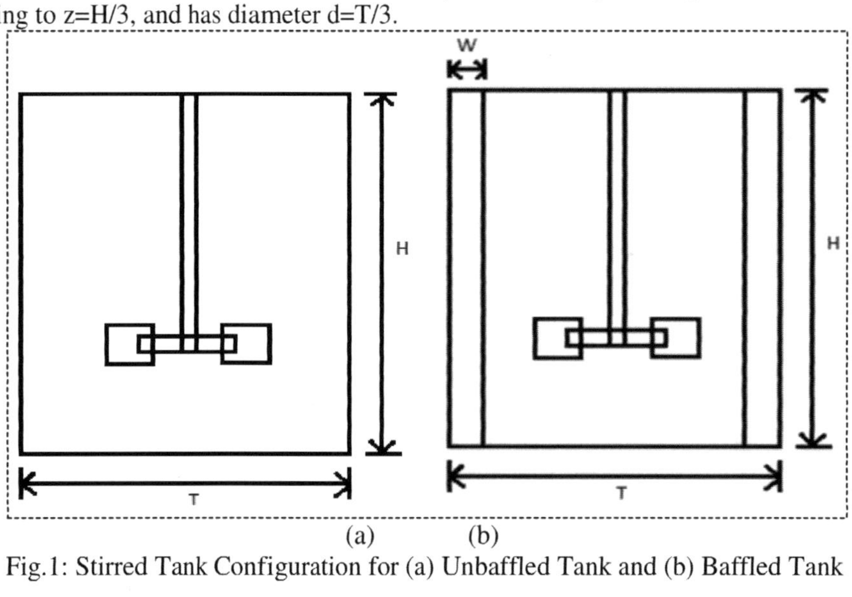

(a) (b)

Fig.1: Stirred Tank Configuration for (a) Unbaffled Tank and (b) Baffled Tank

The types of impeller i.e., Rushton and Curve blade used in this study are presented in Fig.2. The length of blade (l = d/4) and width of blade (b = d/5) is provided. The number of blade used is six and curvature of the Curve blade is provided as 140^0.

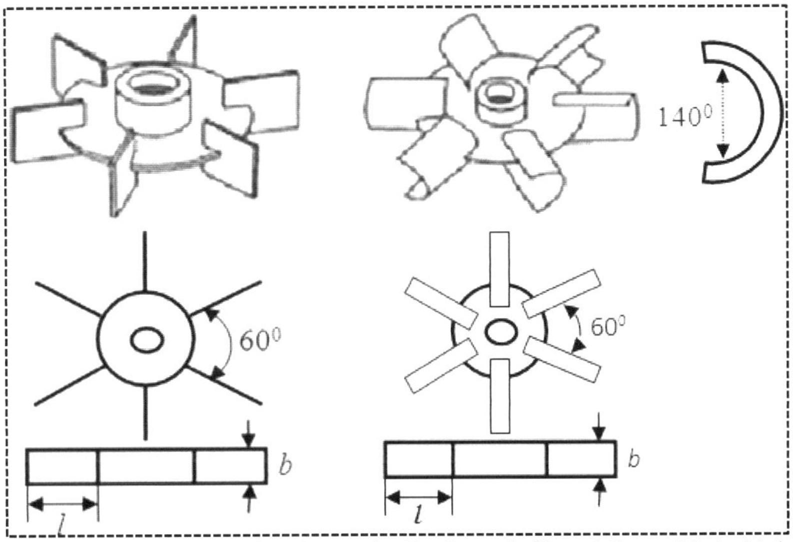

Fig.2: Impeller types used in the study. Rushton (left) and Curve blade (right)

3. Numerical Simulation

In CFD simulation two main equations are involved as continuity and momentum equation. These equations describe the motion of fluid substances and are solved by iteration process in CFD simulation. In the main momentum equation, one more term called Reynolds stresses (velocity component) by averaging the mean and fluctuating velocity is added, and the new equation is known as RANS (Reynolds Averaged Navier-Stokes) equations. To solve this Reynolds stresses, several turbulence models (k-ε; k-ω; Large eddy simulation, LES; Reynolds stress model, RSM; etc.) are available based on certain assumptions. Therefore, the basic theoretical fundamentals and equations (ANSYS, 2017) involved in the numerical simulation using CFD techniques will be presented in this section. Basic numerical equations solved in the CFD simulation are categorized as governing equations and turbulence model equation. The other numerical models & criteria to be adopted such as mesh generation (impeller model, discretization scheme), convergence criteria, boundary condition, etc. is also presented in this section.

3.1 Governing Equations

3.1.1 The Continuity Equation. Continuity equation in conservation form obtained by applying the principle of conservation of mass on a control volume fixed in space and described in differential form is:

$$\frac{\partial \rho}{\partial t} + \nabla.(\rho\, \vec{u}) = 0 \qquad (1)$$

Where, $\vec{u}$ is the flow velocity at a point on the control surface, and ρ is the density of fluid. For incompressible flows, Eq. (1) is simplified to

$$\nabla.\vec{u} = 0 \qquad (2)$$

3.1.2 The Momentum Equation. The conservation form of momentum equation by the application of Newton's second law of motion to a fixed fluid element is described by three scalar equations corresponding to x, y and z directions and for viscous flows they are termed as Navier – Stokes equation. The conservation form is described as:

$$[\frac{\partial u}{\partial t} + (\overrightarrow{u}.\nabla)\vec{u}\,] = -\nabla p + \nabla.\bar{\bar{\tau}} + \rho\overrightarrow{f} \qquad (3)$$

Here, $\nabla.\overline{\overline{\tau}}$ is the viscous stress tensor and is the$\overrightarrow{f}$ body force per unit mass. The viscous stress tensor for Newtonian fluids is given by$\nabla.\overline{\overline{\tau}}$ (in tensor notation):

The Navier-Stokes equation in terms of three scalar components can be described as:

X-component of the momentum equation:

$$\frac{\partial(\rho u_x)}{\partial x} + \nabla.(\rho u_x \overrightarrow{u}) = -\frac{\partial p}{\partial x} + \frac{\partial \tau_{xx}}{\partial x} + \frac{\partial \tau_{yx}}{\partial y} + \frac{\partial \tau_{zx}}{\partial z} + \rho f_x \quad (4)$$

Y-component of the momentum equation:

$$\frac{\partial(\rho u_y)}{\partial x} + \nabla.(\rho u_y \overrightarrow{u}) = -\frac{\partial p}{\partial y} + \frac{\partial \tau_{xy}}{\partial x} + \frac{\partial \tau_{yx}}{\partial y} + \frac{\partial \tau_{zy}}{\partial z} + \rho f_y \quad (5)$$

Z-component of the momentum equation:

$$\frac{\partial(\rho u_z)}{\partial x} + \nabla.(\rho u_z \overrightarrow{u}) = -\frac{\partial p}{\partial z} + \frac{\partial \tau_{xz}}{\partial x} + \frac{\partial \tau_{yz}}{\partial y} + \frac{\partial \tau_{zz}}{\partial z} + \rho f_z \quad (6)$$

The equations (4, 5 and 6) are the Navier-Stokes equations in conservation form. For incompressible flows the second term of the viscous stress tensor given in Eq. (4) is zero due to the incompressibility constraint given in Eq. (2). For constant viscosity, the Navier-Stokes equation for incompressible flows can be written as:

$$\rho\left[\frac{\partial \vec{u}}{\partial t} + (\bar{u}.\vec{v})\bar{u}\right] = -\nabla p + \mu\nabla^2\vec{u} + \rho\vec{f} \quad (7)$$

3.2. Turbulence Model

In this study *k-ε* turbulence model is used. The *k-ε* model is one of a family of two-equation models, for which two additional transport equations must be solved in order to compute the Reynolds stresses. (Zero- and one-equation models also exist, but are not commonly used in mixing applications.) It is a robust model, meaning that it is computationally stable, even in the presence of other more complex physics. It is applicable to a wide variety of turbulent flows, and has served the fluid modelling community for many years. It is semi-empirical, based in large part on observations of mostly high Reynolds number flows. The two transport equations (Eq.8 & 9) that need to be solved for this model are for the kinetic energy of turbulence, *k*, and the rate of dissipation of turbulence, *ε*.

$$\frac{\partial(\rho k)}{\partial t} + \frac{\partial}{\partial xi}(\rho u_i k) = \frac{\partial}{\partial xi} + \left(\mu + \frac{\mu_t}{\sigma_k}\right)\frac{\partial k}{\partial x_i} + G_k - \rho\varepsilon \quad (8)$$

$$\frac{\partial(\rho \varepsilon)}{\partial t} + \frac{\partial}{\partial xi}(\rho u_i \varepsilon) = \frac{\partial}{\partial xi}\left(\mu + \frac{\mu_t}{\sigma_\varepsilon}\right)\frac{\partial \varepsilon}{\partial x_i} + C_1\frac{\varepsilon}{k}G_k + C_2\rho\frac{\varepsilon^2}{k} \quad (9)$$

The quantities C_1, C_2, σ_k and σ_ε are empirical constants. The quantity G_k appearing in both equations is a generation term for turbulence. It contains products of velocity gradients, and also depends on the turbulent viscosity:

$$G_k = \mu_t \left(\frac{\partial u_i}{\partial x_j} + \frac{\partial u_j}{\partial x_i} \right) \frac{\partial u_j}{\partial x_i} \tag{10}$$

$$\mu_t = \rho C_\mu \frac{k^2}{\varepsilon} \tag{11}$$

To summarize the solution process for the k-ε model, transport equations are solved for the turbulence kinetic energy and dissipation rate. The solutions for k and ε are used to compute the turbulent viscosity, μ_t. using the results for μ_t and k, the Reynolds stresses can be computed for substitution into the momentum equations. Once the momentum equations have been solved, the new velocity components are used to update the turbulence generation term, G_k, and the process is repeated.

3.3. Mesh Generation

The purpose of meshing was to divaricate the entire configurational dominion into small control volumes and then the mathematical equations were employed by making use of the computer numerical calculations. For simulation purpose, the mesh was divided into two main region, fluid region of impeller and region of impeller rotation and both were establish as a distinct interacting fluid dominion. Fine meshing was employed so as to increase the stability of the model. CFD code fluent was employed for the purpose of numerical simulation. This CFD technique is based on Navier-Stokes equations which comprises of continuity, energy and momentum equations. In all cases, we used tetrahedral elements for meshing. A total grid size of 214200 (r×θ×z: 25×168×51) was used for the full dominion of the tank accuracy of the computation (Fig.3).3D simulation was done so as to model the flow accurately. For modelling the motion of impeller MRF model was employed. First order upwind discretization scheme is adopted in this study.

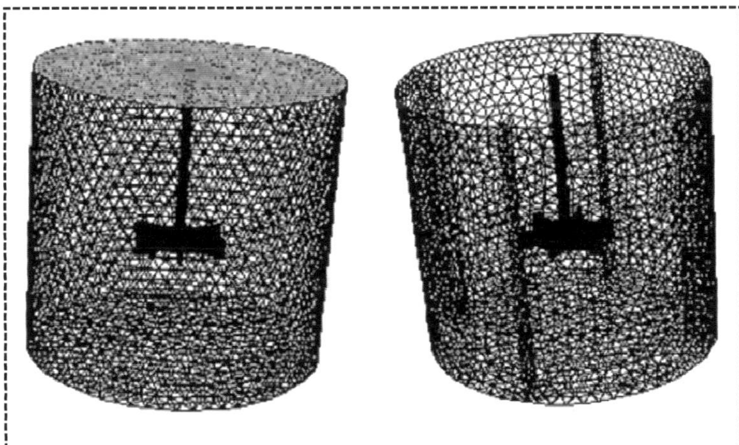

Fig.3: Generated grid for Unbaffled Tank (left) and Baffled Tank (right)

3.4. Boundary conditions

For the unbaffled tanks both Rushton and curve blade the entire computational region was reduced to 60^0,consisting of one blade while for other two vessels which originally consist of six blade and four baffles, the computational region was reduced to 180^0, consisting of 3 blade and 2 baffles. Now there are two types of impeller rotational model multiple reference frame (MRF) and Sliding mesh model. For both the types the entire solution domain is divided into two zones, inner impeller region and outer baffle region. For MRF model, the impeller is consider as rotating reference frame while the outer baffle region is consider as stationary reference frame. It involves steady state calculations as the solution progresses. In sliding mesh model the impeller region is allow to slide over the outer region in discrete time step. In this study MRF model is applied for impeller rotation model.

3.5. Convergence criteria

In this study *k-ε* turbulence model is used for predicting the nature of flow. While normalization and scaling can be done in number of ways, it is the change in the normalized residual that is important in evaluating the rate of convergence. In this study, the simulation is considered to be converged when the residual falls below10^{-3} and steady state of the velocity profile is reached.

4. Numerical Results

The simulated three dimensional mean velocities (radial, tangential and axial) normalized by maximum velocity (U_{tip} = ND where, N is the impeller speed in revolution per sec and D is the impeller diameter in metre) will be discussed in this section. Velocities were measured at a specific spatial location in the stirred tank and compared for baffled and unbaffled condition for Rushton and Curve blade impeller. The predicted results were also compared with experimental results published in literature.

4.1 Axial Profile of Radial Velocity

The normalized radial velocity (U_{radial}/u_{tip}) with respect to normalized tank height (z/H) comparing between baffled and unbaffled system for Rushton and Curve blade impeller is shown in Fig.4(a&b).

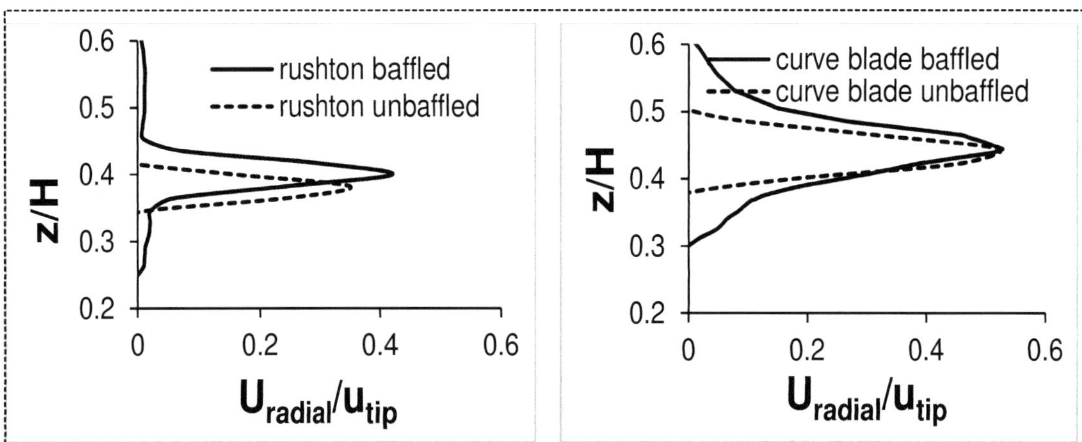

(a) (b)

Fig.4: (a) Radial velocity for Rushton impeller and (b) Radial velocity for Curve blade impeller

In all the four cases, it is found that the radial velocity was maximum in the discharge area of the impeller and then decreases gradually. The radial velocity was found to be highest in baffle tank for both, Rushton and Curve blade impeller. The negative velocity represent that the fluid flow is drawn through the movement of impeller. Specifically in the upper zone it was found that the radial velocity was very low in all the four cases. This proves that the presence of baffle does not impact the radial velocity at the top of tank.

4.2 Axial Profile of Tangential Velocity

Fig.5(a&b) shows the axial profiles of the tangential component of the velocity on the entire height of the tank. In case of Rushton impeller it was found that tangential component of velocity was found to be maximum from z/H=0.27 to z/H=0.4 whereas in case of Curve blade the tangential velocity was found to be maximum from z/H=0.3 to z/H=0.5. A recirculation zone was found to be present on both the sides of impeller in all the four cases. There was sudden decrease of velocity to zero in the unbaffled tank for both the type of impeller which signifies that the movement of fluid was less profound in the unbaffled tank as compare to the baffle tank. Moreover in the unbaffled tank the tangential velocity was found to be most significant than other velocity for both the type of impellers. The tangential velocity was less in baffled tank than unbaffled tank for both the type of impeller which prevents the formation of vortex around the shaft .Hence it was concluded that unlike the radial component the tangential velocity is affected by the presence of baffles.

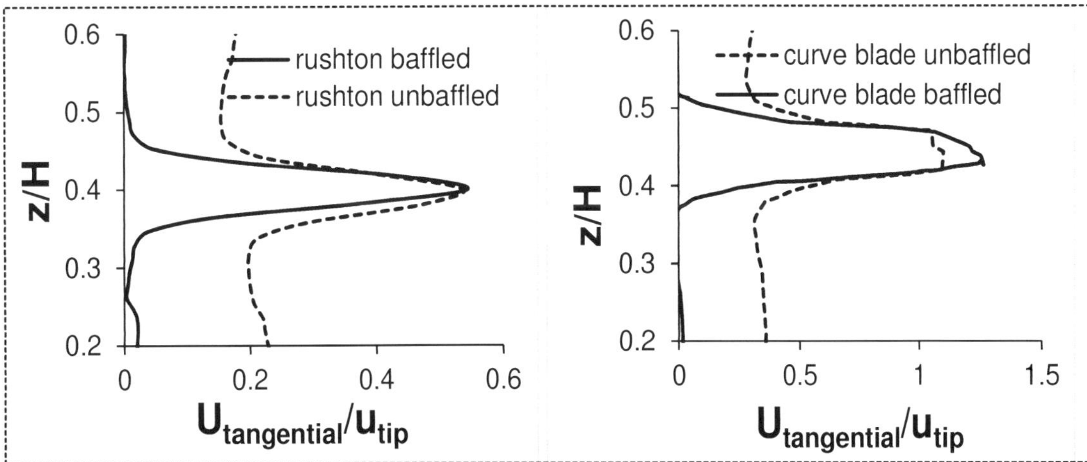

(a) (b)

Fig.5: (a) Tangential velocity for Rushton impeller and (b) curve blade impeller

4.3 Axial Profile of Axial Velocity

Fig.6(a&b) shows the axial profiles of the axial component of the velocity on the height of the tank.There was a great resemblance in all the four cases. A recirculation zone was found to be present on both the sides of impeller. The movement of fluid was found to be more profound in the baffle tank as compare to the unbaffled tank .The increase in the upward axial jet was due to the presence of baffles .At the topmost part of the tank the negative value of the axial velocity indicates that flow is drawn at the shaft to the power of impeller. Thus it was concluded that the presence of baffles intensify the axial jet in both the impellers.

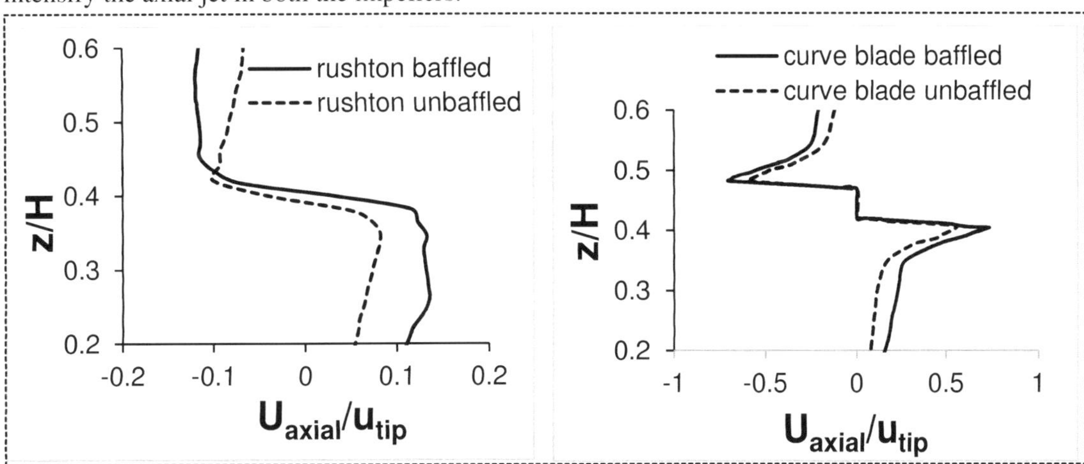

(a) (b)

Fig.6:(a) Axial velocity plot for Rushton impeller and (b) curve blade impeller

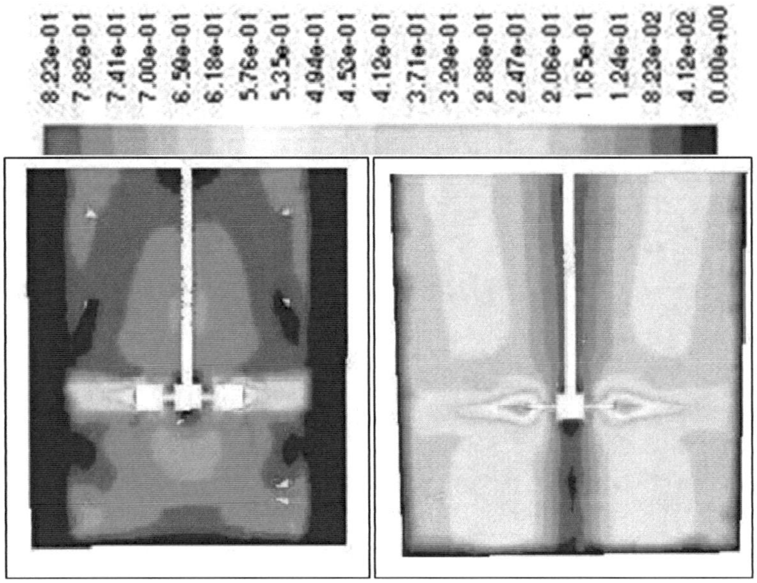

(a) Unbaffled (b) Baffled

Fig.7: Distribution of the mean velocity (contours) when y=0 for (a) unbaffled and (b) baffled system

Fig.7 shows the distribution of the mean velocity in the r-z plane containing the blades. It is observed that the appearance of a strong centrifugal flow from the blades that reached up to the sidewall of the tank placed in front of the radial jet. In a cylindrical tank, the flow generated by the rotation of the Rushton turbine is very low. The radial jet does not reach the lateral walls. As a result, the presence of the baffles intensified the radial jet which reached the lateral surface which, in turn, caused two axial jets, one up and one down. In the baffled stirred tank, the axial jet was found to be the most significant. This upward jet brought up the recirculation loop extended in the upper area of the tank thereby feeding back to the turbine. We noted a great similarity in the lower area of the second configuration and that of the, presumably because the systems is baffled in the lower zone. Moreover, we noticed a recirculation zone which arose to power the Rushton turbine. But in the upper zone, the upward axial jet dropped sharply in the absence of the baffles in the top of the tank. The recirculation loop in the upper zone was less extensive. Thus, we concluded that the use of baffles on the height of the tank increases the axial 19 jets on either side of the turbine, and furthermore, promotes the circulation of the fluid throughout the volume of the tank. The predicted flows reveal that the Rushton turbine produces a radial flow, which upon impingement on the reactor walls deflects upwards towards the liquid surface. The stream then returns to the stirrer, thus forming a recirculation zone. The principal difference between the flow in an unbaffled stirred tank and that in a baffled tank is the rotation of the liquid, which usually acts as a combined vortex consisting of an inner core of forced-vortex (or solid-body rotation) and an outer region of free-vortex motion resulting in the deformation of the liquid free-surface. In unbaffled agitated vessels, a large vortex is formed because of the free headspace resulting in the deformation of the liquid surface. This has a significant effect on flow and mixing processes.

4.4 Impeller Efficiencies based on Power Consumption

For describing the effectiveness of a mixing system the power consumption is a very significant framework. Fig.8 represents power efficiency for all the four type of tank configuration.

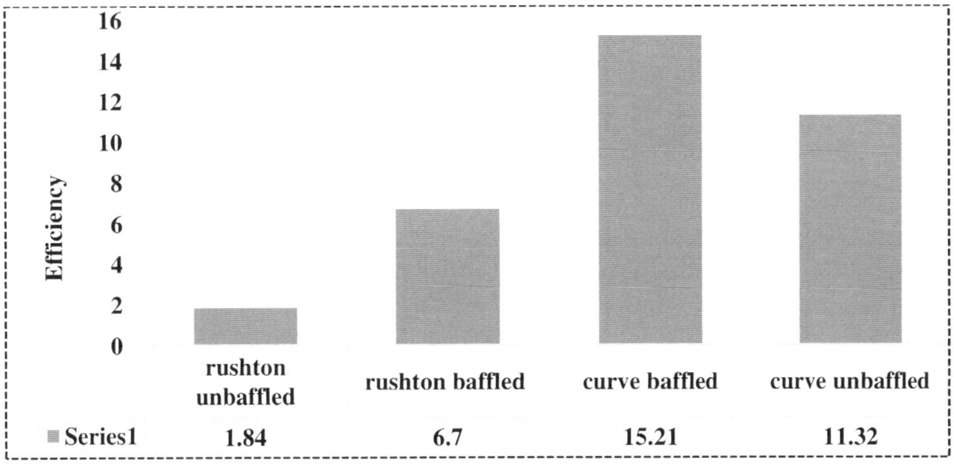

Fig.8: Power efficiency for all the four type of tank.

In this study the efficiency E, of the impeller has been calculated by including two flow factors impeller flow number N_d, impeller power number N_p as:

$$E = N_d^3/N_p \quad (12)$$
$$N_P = P/n^3d^5\rho \quad (13)$$
$$N_d = Q/nd^3 \quad (14)$$

Where, P is the power consumption of the tank, n is the impeller rotational speed, d is the impeller diameter and ρ is the fluid density. For both the type of blades, the power consumption of a baffle tank was found to be greater than the unbaffled tank. Also the efficiency of the baffled tank for both Rushton and Curve blade impeller was found to be better than unbaffled tank .This is due to the presence of low velocity gradient in the baffled tank as compare to the un-baffled tank.

4.5. Comparison with Experimental Results

Fig.9 (a&b) shows the axial profiles of the radial and tangential velocity components in the radial position defined by r/D = 0.185 in the baffled stirred tank equipped with the Rushton impeller.

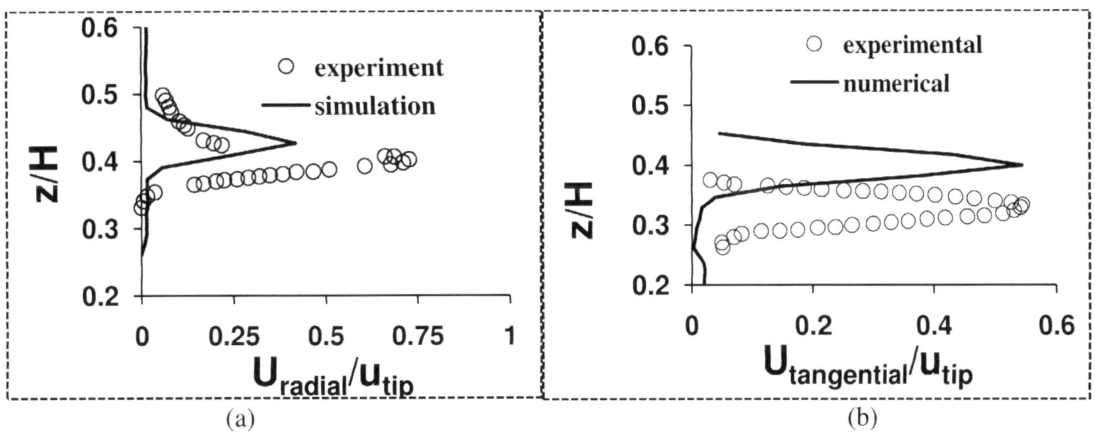

Fig.9: (a) Axial profiles of the radial velocity and (b) Axial profiles of tangential velocity

Mean fluid components of an aeration tank should be speculated precisely by a CFD model. However *k-ε* turbulence model has been found to under or over project turbulence. In the present work it has been found that there is a diversion between numerical and experimental data [35] particularly underneath the impeller. In this study the experimental data comprises of 10 data sets from which a continuous graph has been derived. Also any contrast between numerical and experimental data in both this and upcoming sections should be comprehend with the characteristics of the experimental data set [35] in mind i.e. results should be analogous rather than exactly alike.

5. Conclusions

The outcome of numerical evaluation dispensing the effect of vessel configuration on the flow structure and turbulent characteristics has been evaluated. The present work examines the impact of baffle and blade on the power consumption of the impeller. For all the four system it has been found that power number is dependent on the length of baffle. Also the use of baffle increases the circulation and enhances the degree of mixing. The outcome predicted also stipulates that the power consumption and efficiency of tank is influenced by the type of blade used. Although the power consumption of baffle tank was higher than the unbaffled tank, the efficiency of baffled system was better than the efficiency of unbaffled system. Overall the efficiency of the Curve blade impeller was found to better than the other configuration.

References

[1] Lane G L *et al* 2005 *Chem. Engg. Sci.***60** 2203
[2] Deglon DA and Meyer CJ 2006 *Miner. Engg.***19** 1059
[3] Gaubert M A *et al* 2006 *Chem. Engg. Proc.***45** 415
[4] Zadghaffari R *et al* 2009 *Computatio. Chem. Engg.* **33** 1240
[5] Ranade V V*et al* 1991 *Chem. Engg. Sci.***46** 1883
[6] Bridgeman J 2012 *Adv. Engg. Soft.***44** 54
[7] Gentric C *et al* 2005 *Chem. Engg. Sci.***60** 2253
[8] Nagata S 1975 Wiley, New York
[9] Glover G M C and Fitzpatrick J J 2007 *Chem. Engg. J.***127** 11
[10] Assirelli M *et al* 2008 *Chem. Engg. Sci.* **63** 35
[11] Aloi L E and Cherry RS 1996 *Chem. Engg. Sci.* **51(9)**1523
[12] Hekmat D *et al* 2007 *Proc. Biochem.* **42** 1649
[13] Rousseaux J M *et al* 2001 *Canadian J. Chem. Engg.***79** 697
[14] Kagoshima M and Mann R 2006 *Chem. Engg. Sci.***61(9)** 2852
[15] Tezura S *et al* 2007 J. *Chem. Tech. Biotech.***82** 672
[16] Galletti C and Brunazzi E 2008 *Chem. Engg. Sci.***63** 4494
[17] Yoshida M *et al* 2008 *J. Fluid Sci. Tech.***3** 282
[18] Hirata Y *et al* 2009 *Chem. Engg. Res. Des.***87(4)** 430
[19] Brucato A *et al* 2010 **Chem. Engg. Sci**. **65** 3001
[20] Wang B *et al* 2012 *Biotech. Adv.* **30** 904
[21] Shekhar S M and Jayanti S 2002 *Trans. IChemE.* **80(A)** 482
[22] Alcamo R *et al* 2005 *Chem. Engg. Sci.***60** 2303
[23] Cokljat D *et al* 2006 *Prog. Comp. Fluid Dyn. Inter. J.***6(1-3)** 168
[24] Nienow A W 1996 *Trans. Inst. Chem. Engineers***74(A)** 417
[25] Yoshida M *et al* 1996 Canadian J. Chem. Engg.**74(1)** 31
[26] Myers K J *et al* 1999 *Chem. Engg. Res. Des.***77** 728
[27] Bakker A 2000 The Online CFM Book
[28] Sardeing R *et al* 2004 *Trans. IChemE***82(A9)** 1161
[29] Albaek M O *et al* 2008 *Chem. Engg. Sci.* **63** 5813
[30] Gimbun J 2009 *Chem. Engg. Sci.***87** 437
[31] Ahmed S U *et al* 2010 *J. Biosci. Bioengg.***109(6)** 588
[32] Suhaili N *et al* 2010 *J. Appl. Sci. Res.***6(3)** 234
[33] Suhaili N *et al* 2011 UMTAS, 146

[34] Van'tRiet K *et al* 1976 *Chem. Engg. Res. Des.***54** 124
[35] Wu H and Patterson G K 1989 *Chem. Engg. Sci.***44** 2207

Challenges in joining of metallic pipes using microwave hybrid heating

D N Gamit[1*] and M K Chudasama[2]

[1]Gujarat Technological University, Ahmadabad, Gujarat, India-382424
[2]Mechanical Engineering Department, Govt Engg College, Dahod, Gujarat-389151
[*]Email: dngamit@gmail.com

Abstract. Microwave hybrid heating (MHH) is a distinct and emerging alternative joining technique used to join metallic materials compared to available conventional joining processes such as FW, TIG, MIG, etc. MHH shows significant characteristics such as selective and volumetric heating, lower processing time, eco-friendly processing against conventional joining. This paper aims to investigate the joining of 316L stainless steel pipes by incorporating micrometric size (45μm) of 316L stainless steel powder in between the faying surfaces of the candidate material. Charcoal as a susceptor material was placed at the joint zone and exposed to microwave energy at 2.45 GHz inside the domestic microwave applicator at 900 W power levels. The challenges and principles involved in joining SS 316 L pipes using MHH have been discussed and initial experimental observations have been analyzed.

1. Introduction

Metallic pipes are widely used to transport fluid of gases at different pressure and temperature in industries such as marine, chemical, food, petroleum, etc. The alloy pipe is used for such applications due to excellent mechanical properties and higher resistance to corrosion at elevated temperatures due to the addition of molybdenum. In manufacturing processes, the fabrication of continuous structure of metallic pipe is not a feasible way hence; the discrete pipes need to be joined using conventional joining techniques such as LBW, TIG, SMAW, etc. Presently, the use of conventional joining technique is well developed in terms of technical aspects, however; the fusion welding of thin-walled SS pipe is not convenient with such high energy processes which damage the joint area largely and deteriorate the mechanical and microstructural properties with the heat input due to the lower conductivity of SS pipe. Secondly, the problem associated with the sound joint formation of thin-walled pipes joint properties will be governed by operator skill and working conditions [1–3]. More processing time and energy consumption can be wasted in frequent rework during the maintenance of pipe joints, hence; industries are searching for a viable and alternative joining process to overcome such problems and join thin-walled SS pipes with good reliability of the joint. Recently, microwave hybrid heating (MHH) techniques emerge as a novel route of joining techniques for metallic material with advantages such as lower heat input, uniform joint heating, eco-friendly joint formation, etc. as compared to conventional joining techniques. Use of microwave energy as an engineering tool in different field of application such as food processing, chemical synthesis, sintering, joining, etc. have been reported by many researchers with unique features such as uniform and selective heating characteristics, eco-friendly and energy-saving process, reduced processing time, better mechanical and metallurgical properties [4–18]. It has been appearing from the literature that microwave energy has the potential to address issues, related to metallic pipe joining. The microwave hybrid heating (MHH) technique was reported first time in the form of Indian pattern in the year 2009 [13]. Later, in the year 2011, Srinath et al have reported the joining of copper plates by introducing copper powder as interface material [14]. The joint of the microwave welded copper plate shows higher joint strength than the strength of the TIG welded joint. The joint of SS-316 plates developed by MHH was reported, higher joint strength, and higher microhardness using the micrometric size of SS-316 powder [15] than Nickel powder [16]. The MS plates [17] and MS pipes [18] were joined using microwave energy by placing nickel powder (40 μm) as a sandwich layer and it was reported significantly joint strength. Similarly, dissimilar metal joining (SS 316 and MS) was also reported by the researcher with better joint strength and microhardness [6-7]. Inconel-625 alloy butt joint was developed using microwave energy by adding micron size nickel powder as interface material. The authors have reported that increase in specimen size also led to increase in microwave exposure time [9]. The available literature shows that the MHH technique is established as a metal joining process. The present paper reports the challenges encountered, basic principles of metallic joint development, and initial experimental results on SS 316L pipe joining using the MHH technique at 2.45 GHz and 900 W.

2. Challenges

Microwave processing of bulk metallic pipe is quite challenging, yet an interesting research area due to the unique heating characteristic of the microwave energy and its interaction with the materials. The literature on microwave joining of pipes is hardly available in this area. Microwave joining of metallic pipes required knowledge and fundamental aspects of microwave processing of materials. Control of MW energy in a particular application is a key issue for processing the material. The challenges encountered using microwave energy during the joining of metallic pipes are shown in Figure 1 and Table.1.

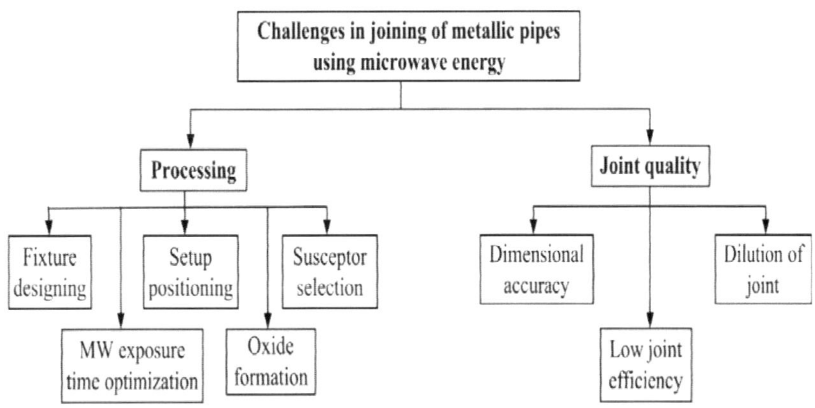

Figure 1: Challenges encountered using microwave energy during the joining of metallic pipes

2.1. Fabrication of fixture

In metal joining using MHH technique, the fixture designing plays most important role as it protect the metal from arcing and act as a masking material. Alumina based fire brick (ceramic material) is used for fixture as it is low dielectric loss substance which absorbs lesser amount of MW energy and shielded metallic pipes. It also supports the microwave energy to intensify at a joint area of the substrate material for selective heating as it holds the susceptor material. Designing of fixture consists of susceptor selection, masking material, the location and position of a fixture in the microwave applicator cavity and orientation of fixtures, etc. The microwave key characteristics such as uniform and selective heating solely depends on selection of fixtures materials whereas joint alignment, minimization of HAZ, application pressure etc. confined to positioning of fixture. The vertical position of the fixture as shown in Figure.2 provides better results due to ease of susceptor distribution around joint zone which promotes more uniform and selecting heating of the interfacing and joint area. In the horizontal positioning of the fixture, the susceptor placing at the joint zone was difficult due to the circular geometry of the substrate. The thickness of the susceptor around the pipe was not maintained similarly hence; the non-uniform heating of the joint zone takes place leading to improper fusion of filler and base material.

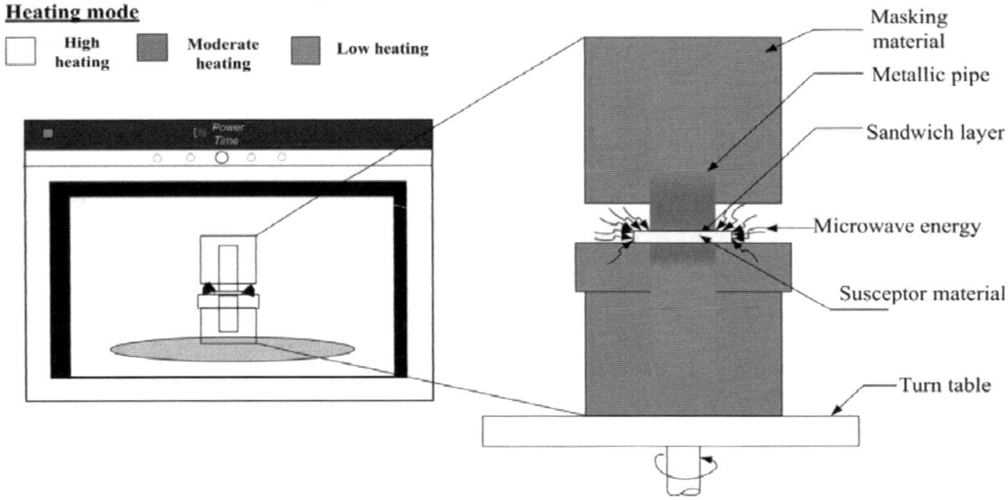

Figure 2: Position of fixture in microwave applicator

2.2 Selection of susceptor materials

Susceptor materials such as charcoal powder, SiC, graphite, etc. plays very crucial role as it acts as a source of heat during MW irradiation. It must be a high dielectric loss substance that could be directly couple with microwave energy at room temperature when expose to microwave radiation and gets heated up very rapidly. Susceptor is placed at the joint area so that the fusion of materials ensure. The selection of susceptor material depends upon the characteristics such as higher heating rate, oxidation prevention, less contamination in the joint zone.

Table 1: Some other challenges in joining of metallic pipes

Basis	**Challenges in metallic pipe joining**
Physics	Inadequate literature for better understanding of the microwave-metal interaction phenomena
Interface materials	Use of optimum size of the powder and quantity of filler materials which maintaining required porosity for better joint strength
Mathematical models	Rarely available mathematical models to simulate behaviour of microwave processing effects in metal based materials
Handbook on Nano-size metal	Unavailability of microwave processing data hand book on nano-size metal

3. Experimentation

3.1 Selection of materials

The use of SS pipe is worldwide due to its excellent mechanical and physical properties. The SS pipe encountered problems while joining with the conventional process due to changes in the microstructural transformation due to higher heat input characteristics of the process. To control such a problem, the process required a higher vacuum which increases the cost of processing. The use of MW energy for joining material doesn't required vacuum as well as not affect the structural transformation of steel. Thus, the SS 316L pipe as shown in Figure 3 was selected as a candidate material for joining using MHH. The chemical composition of the SS 316L pipe is shown in Figure 3. Experimental trials were carried out in the domestic microwave applicators (Model: make: LG) with maximum power 900 W at 2.45 GHz to join commercially available SS 316L pipes having inner and outer diameter are 10 mm and 14 mm with pipes length 30 mm respectively.

Chemical composition (Wt %) of the SS 316L pipe

C	Mn	Si	Cr	Ni	Mo	Fe
0.015	0.97	0.390	16.56	9.95	1.97	Balance

Chemical composition (Wt %) of the SS 316L powder

C	Mn	Si	Cr	Ni	Mo	Fe
0.05	0.18	1.13	17.90	12.24	2.90	Balance

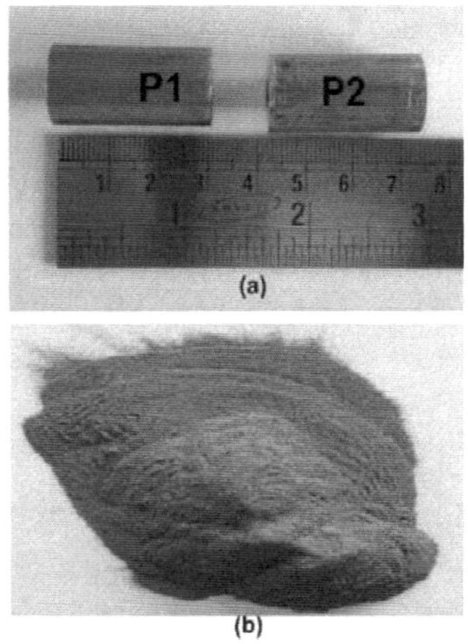

Figure 3: As received (a) SS 316L pipe (b) SS 316L powder as interface material

The SS powder of micrometric size (45 μm) was used as an interface material in the butt joint. Table.3 show the chemical composition of as received SS 316L powder. An epoxy resin (Bisphenol - A based) was used as a binder with SS powder and forms slurry which was placed at faying surfaces of SS steel pipes. A susceptor (a high dielectric loss material) was used for hybrid heating of the pipe area to be joined. Charcoal was used as a susceptor material. Ceramic was used to develop a fixture for selective heating of the joint area and hiding the metallic pipes during the joining process.

3.2 Development of joint

The slurry was prepared by adding approximately 10% of the binder (epoxy resin) in SS powder and was uniformly placed over entire surfaces of pipes to be joined. At the initial stage, slurry helps to bind the substrates and maintained the alignment of it. The binder used in the slurry also absorbed microwave energy and it gets evaporated above 300°C. There was no additional effect of epoxy resin observed in the formation of the pipe joint. Bulk metallic pipes reflect the microwave at room temperature due to lower skin depth; therefore, the masking material was used to cover the metallic pipes to avoid this problem. A susceptor that absorbs MW energy rapidly and acts as a source of heat when exposed to microwave irradiation. Due to the heating of the susceptor, the high heat was generated near the area to be joined. Bulk metals cannot directly coupled with microwave energy at room temperature hence, initial heating of interfacing materials and faying surfaces of the substrate taking place due to the transfer of heat from susceptor through conduction. Beyond critical temperature, metal directly absorb microwave energy and heating of metal taking place. Figure 4. Illustrate the schematic diagram of the experimental setup of the joining of SS pipes by MHH.

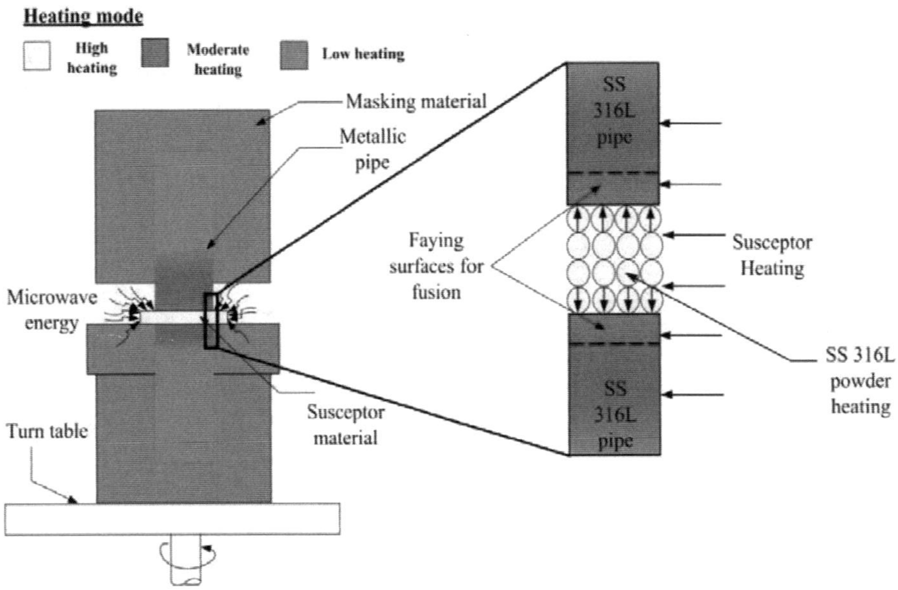

Figure 4: Schematic diagram of microwave joining of SS 316L pipes

After initial heating, metallic pipes reach critical temperature and start absorbing microwave energy resulting in the melting of the sandwich layer, and faying surfaces of pipes led to fusion and excellent bonding of candidate materials. The experimentation was carried out in a microwave applicator at the environmental condition. The details of microwave processing parameters are summarized in Table 2

.Table 2: Microwave processing parameters

Parameters	Description
Microwave frequency	2.45 GHz
Candidate material	SS 316L pipes
Interface material	SS 316L powder
Susceptor material	Charcoal powder
Exposure time	840-900 s
Exposure power	900 W

4. Initial experimental results and discussion

The experiments were carried out using charcoal powder as a susceptor material and the initial observations of joining of SS 316L pipes using MHH are shown in Table 3. The microwave hybrid heating technique using different grade charcoal powder with different exposure times at 900 W powers. Activated charcoal powder (fine powder) with exposure time increases up to 900 s causes melting and sintering of interfacing powder with the faying surfaces of the SS pipes. The problem associated with the fine charcoal powder was the dispositioning of the powder after microwave processing. Due to light in weight, fine powder was not stick around the joint zone leading to improper heating of the joint as shown in Figure 5(c). The melting of sandwich powder was observed up to 900 s exposure times; however, no melting of the base metals was achieved. This can be attributed to the formation of an oxide layer at the joint zone which requires more heating to melt the faying surfaces of pipes. The hard coke powder gives the best result due to the proper covering of the joint zone even after microwave processing. It allows a good heating rate resulting in the melting of the interface powder and no oxidation was observed.

Table 3: Effect of susceptor materials and exposure time on joining of SS pipes at 900W power

Sr. No	Susceptor	Exposure time (s)	Results
1	fine Charcoal powder	540	No joining
2		600	No joining
3		780	No joining
4		900	Charcoal powder not sticks at the joint zone. Sintering of interface material with faying surface
1	Hard cock Charcoal powder	600	No joining
2		840-900	Melting of interface powder was observed. Joining of SS pipes was observed

The exposure time below 840 s was insufficient to complete heating and melting of interfacing powder and the base metals with high density fire brick. However, appropriate fusion of base metal and interface powder was observed at 840 to 900 s as shown in Figure 6 (c). The rapid response, eco-friendly nature, and less time-consuming characteristics of microwave energy have attracted researchers to develop microwave processed products for industrial usages. The better understanding of process physics, mathematical models of the process, and control over environmental conditions in microwave applicators will help in achieving better quality, higher efficiency, and robust process design.

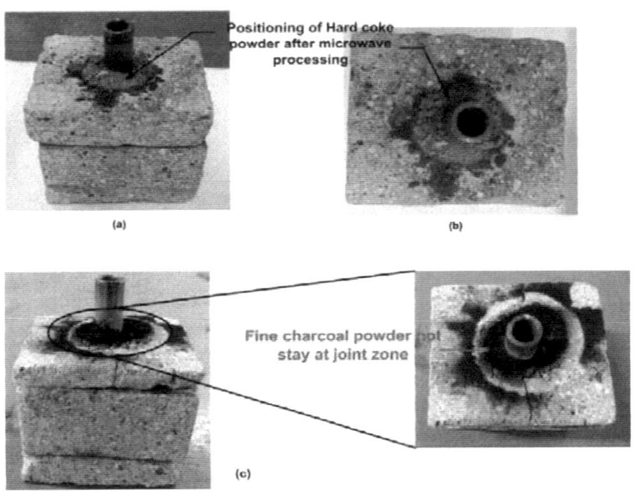

Figure 5: Location of (a) & (b) hard coke charcoal powder (c) fine charcoal powder.

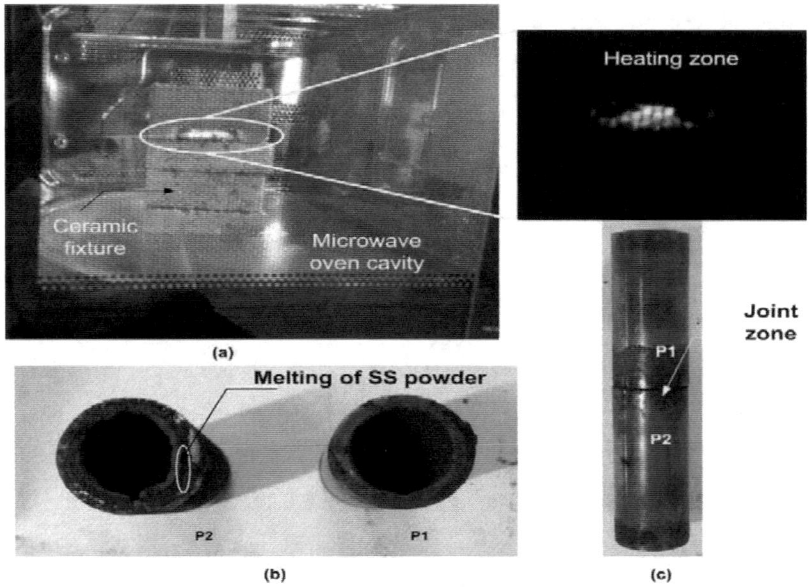

Figure 6: (a) Positioning of fixture inside MW oven. (b) Melting of SS powder was observed at 840 s Exposure time (c) Developed SS joint

5. Limitation

There are some limitations on which researchers have to deal with during the processing of metallic materials using microwave energy. The designing of a foolproof ceramic fixture is very essential as improper designing of fixture responsible for thermal damaging of the substrate; large size products cannot be processed due to fixed-size domestic microwave oven available. Processing of large size products required large microwave applicators which required higher investment. Microwave processing of metal is a newly developed technique due to which very little experimental data available, therefore it limits the flexibility of modeling and simulation of microwave processing and also leads to poor repeatability of experimental results that restrict the use of this process in Industrial applications. Microwave leakage is very harmful to humans hence precaution and safety measures should be followed during microwave processing.

6. Conclusions

The microwave hybrid heating (MHH) technique was used to join SS 316L pipes by incorporating 45 μm size of SS powder at 900 W power in a microwave applicator. The initial observations show that the joining of the pipe was obtained at 840-900 s exposure time with hard coke charcoal powder. The microwave joining process has been investigated by various researchers; however, the process is yet to be industrialized. The rigorous study of process physics and mathematical modeling tuned the process and used as an alternative source of the conventional pipe joining processes.

Acknowledgement
The authors would like to acknowledge the help and guidance provided by Dr. Radharaman Mishra, IIT Roorkee and Dr. Apurbbakumar Sharma, Professor in Mechanical and Industrial Engineering Department, Indian Institute of Technology Roorkee, Uttarakhand, India in the article.

References

[1] Groche P, Wohletz S, Brenneis M, Pabst C and Resch F 2014 *J. Mater. Process. Technol.* **214,** 1972
[2] Silva C M, Nielsen C V, Alves L M and Martins P A 2015 *J. Cleaner. Prod.* **87,** 777
[3] Yapp D and Blackman S A 2004 *J. Braz. Soc. of Mech. Sci. & Eng.* **26,** 89
[4] Clark D K, Folz D C and Jon K 2000 *Mate.r Sci. Eng. A.* **287,** 153
[5] Thostenson E T and Chou T W 1999 *Compos. Part A Appl. Sci. Manuf.* **30,** 1055
[6] Srinath M S, Sharma A K and Kumar P 2011 *J. Manuf. Process.***13,** 141
[7] Bansal A, Sharma A K, Kumar P and Das S 2016 *Proc. Inst. Mech. Eng. Part B J. Eng. Manuf.* **230,** 439
[8] Gupta P and Kumar S 2014 *Mater. Manuf. Process.* **29,** 910
[9] Badiger R I, Narendranath S and Srinath M S 2015 *J. Manuf. Process.***18,** 117
[10] Gamit D N and Chudasama M K 2020 *J. Mech. Eng. Sci.* **14,** *6770*
[11] Mishra R R and Sharma A K 2016 *Compos. Part A Appl. Sci. Manuf.* **81,** 78
[12] Sharma A K and Mishra R R 2018 *Mater. Sci. Technol. (United Kingdom).* **34,** 123
[13] Sharma A K, Srinath M S and Kumar P 2009 Microwave joining of metallic materials, Indian Patent.1994
[14] Srinath M S, Sharma A K and Kumar P 2011 *Mater. Des.***32,** 2685
[15] Bansal A, Sharma A K, Kumar P and Das S 2011 *Mater. Charact.* **91,** 34
[16] Srinath M S, Sharma A K and Kumar P 2010 *Proc. Inst. Mech. Eng. Part B J. Eng. Manuf.* **9,** 1083
[17] Bansal A, Sharma A K, Kumar P and Das S 2012 *Adv. Mater. Res.* **585,** 465
[18] Gamit D N, Mishra R R and Sharma A K 2017 *J. Manuf. Process.* **27,** 158

IOP Conf. Series: Materials Science and Engineering 1004 (2020) 012008 doi:10.1088/1757-899X/1004/1/012008

Effectiveness of Seismically Resistant Braced Frames to Blast Loading

Akshay Kapoor[1*] and Saraswati Setia[2]

[1]Post Graduate Student, National Institute of Technology, Kurukshetra, Haryana, India
[2]Professor, National Institute of Technology, Kurukshetra, Haryana, India
[*]Email: akskap.96@gmail.com

Abstract. The increment in the tally of terrorist radicalisation from the past time has exhibited that the consequence of blast loads on structures is a crucial matter that should be taken into cogitation. This paper considers the effect of blast load on seismically designed framing systems namely Moment Resisting Frame (MRF), Concentric Braced Frame (CBF), Eccentric Braced Frame (EBF) and Buckling Restraint Braced Frame (BRBF). The bracings are one of the efficient lateral load resisting systems and their behaviour is observed under the blast loading. The burst load is estimated to be an unconfined surface burst explosion on front face of the structure. Comparative analysis of storey displacement among different framing systems is performed. Failure of members is studied through the analysis of plastic hinge formation. The results indicate that braced frames perform better when exposed to such kind of impact load. Buckling Restraint brace (BRB) exhibited higher level of resistance to blast loading due to its ability to yield in both compression and tension.

1. Introduction

Over the last few decades, accidental explosions and violent attacks due to radicals have brought urgency in inquest in the field of response of structures that are exposed to blare produced from burst of explosives. Deterioration of structures amplified with the deprivation of mortal lives occur when a bomb explosion takes place within or surrounding a building. Major attacks happened in the past were the Oklahoma bombing (1995), car bomb explosion outside the main terminal of Glasgow international airport, Britain (2007), Zaveri Bazaar bombing case in Mumbai (2007), bomb explosion in front of Aleppo Hotel in Syria (2014) and many more.

This paper deals with investigating the behaviour of seismically designed frames having the same configuration subjected to blast loading. The seismically designed frames include Moment Resisting Frame (MRF), Concentric Braced Frame (CBF), Eccentric Braced Frame (EBF) and Buckling Restraint Braced Frame (BRBF). The behaviour is investigated using generalised Finite Element Method (FEM) based software tool i.e., ETABS v16.2.1. The blast load is assumed to be an unconfined surface burst detonated from 15 m from the front face of the building. Adeli et al. [1] used the applied element method to analyse the different steel framing systems and concluded that EBF had fewer number of failed members when subjected to shock loading.

1.1. Buckling Restraint Braced Frame (BRBF)

BRBFs are one of the advanced systems of lateral load resisting systems used in the modern structures [2]. The geometric configuration of BRBF is similar to that of the Concentric Braced Frame (CBF). The difference lies in the composition of BRB itself. A schematic diagram describing the different components of BRB is shown in figure 1.

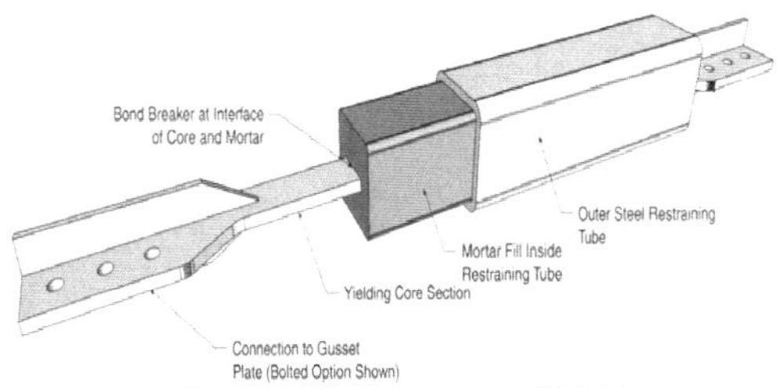

Figure 1: Components of BRB

2. Blast Load Phenomena

A sudden release in energy creates pressure disturbance which is known as a blast. This pressure disturbance is created due to rapid expansion of an energetic material. The phenomena signifying that the blast wave is radiated from the origin of fulmination is expressed in figure 2. The pressure amplitude of blast load rises to higher degree and the eternity of the load from burst also inflates with the propagation of blast wave from the origin [3]. The creation of vacuity at the origin takes place and also the flip-flop of advancement of gas occurs due to overexpansion at the centre of the blast. Due to outward expansion of negative pressure region, a negative pressure is created that is below the ambient pressure. The positive phase is trailed by this negative phase. The pressure generated during negative phase has lower magnitude than pressure generated during positive phase but it has prolonged duration than the latter. Generally, out of the two phases of burst, the negative phase is less important so it can be omitted.

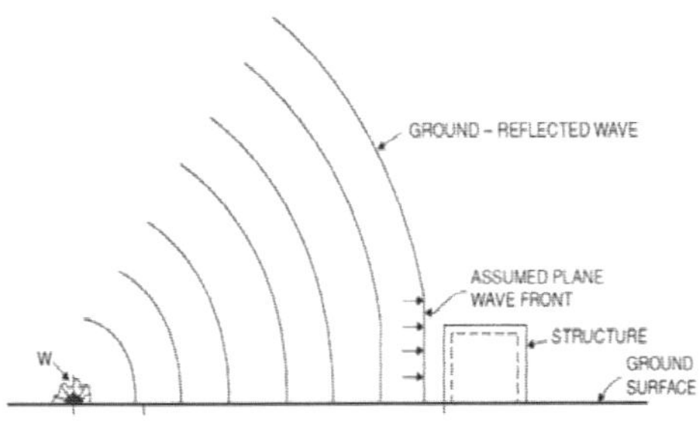

Figure 2: Propagating Blast Wave

2.1. *Blast Wave Characteristics*

The propagation of the reaction front is supersonically more rapid than the speed of sound in a detonation process. This wave propagated from the blast is called as a shock wave. Blast wave has a catastrophic jump in pressure. The pressure disturbance at any end from the blast point, has the shape shown in figure 3. At time t_A, the arrival of shock front at a given location gives rise to the incident pressure with its peak value, P_{so}, followed by its perish to the clime value in duration called as the positive phase duration. After that there is a negative lapse with a time lapse t_o^- which is more than the positive lapse and it is designated by the negative pressure which has a greatest value of P_{so}^- with the flip-flop of flow of particles. The magnitude of the negative pressure is well below the magnitude of ambient pressure. In design, generally this negative lapse is of less significance than the positive lapse, and its magnitude P_s^- is, in all cases, less than the clime atmospheric pressure p_o. For blast wave, the area resultant of the integration under the pressure v/s time curve is defined as the incident impulse density which is described as i_s^- for negative lapse and i_s for positive lapse.

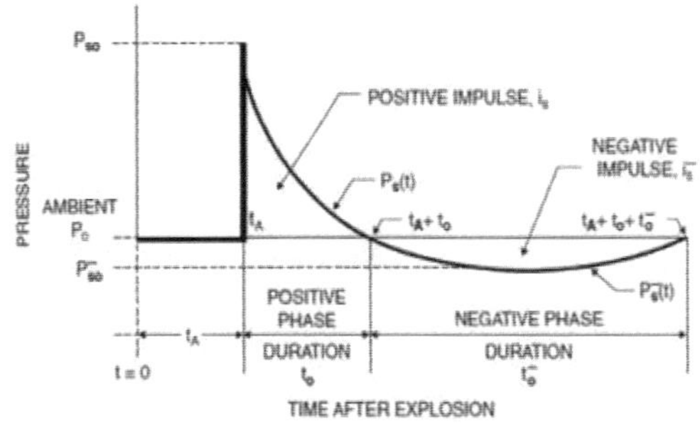

Figure 3: Shock Wave Propagation

3. Modelling

Four 7-storey Reinforced Cement Concrete (RCC) frames including MRF, CBF, EBF and Buckling Restraint Braced Frame (BRBF) are modelled using ETABS v16.2.1 software. The geometry of the bracings is kept as inverted – V shape. The 3-D view of the considered frame is shown in the figure 4.

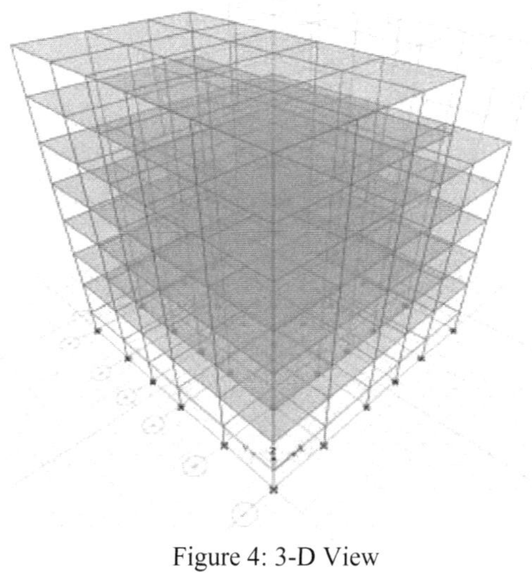

Figure 4: 3-D View

3.1. Blast Wave Characteristics

The frame consists of five bays both in x and y direction. Length of three bays is 4m each and that of two bays is 3m each. Hence the plan dimension of building is 18m x 18m. Depth of foundation is kept 2m. Height of each storey is 3m with total height of 21m above plinth level. Dimensions of different structural components of frame are shown in table 1.

Table 1: Model Description

Structural Elements	Dimensions
Column	350mm x 450mm

Beam		300mm x 300mm
Plinth beam		250mm x 250mm
Channel section		ISMC 300
BRB	Tube	200mm x 200mm
	Core plate	80mm x 8mm
Slab		150 mm thick
Brick wall		230 mm thick

3.2. *Structural Details*

The design considerations of Buckling Restrained Braces (BRB) are taken in accordance with the provisions from National Institute of Standards and technology (NIST) brief no. 11, which is a guide for practicing engineers issued by U.S. Department of Commerce under National Earthquake Hazards Reduction Program (NEHRP) [4]. The strength parameters of different components of the considered frame are shown in table 2.

Table 2: Structural Details

Grade of Concrete		M35
Grade of Rebar		Fe 500
Grade of Steel		Fe 250
Grade of BRB components	Steel tube	290 MPa
	Core plate	345 MPa

Length of yielding core of BRB is considered as three-fifth brace length. Area of flat steel plate to be used as core plate is taken as 6.4 cm .

4. Blast Load Calculation

The blast parameters due to the detonation of charge weight of 100 kg (Trinitrotoluene) at a standoff expanse of 15 m from the front face of the building are calculated with reference to IS 4991: 1968 (reaffirmed 2013).

4.1. *Front Face Loading*

At any time t, the net pressure that occurs on the fore front of building is :
The reflected pressure p_r or ($p_s + C_d$ q), whichever is greater (1)

C_d = drag coefficient
p_r = The value of overpressure (reflected) falling from the max. value p_{ro} to overpressure in duration t_c is described as:
t_c = 3S/U or t_d whichever is less

Where,
S = H (height of building) or B/2 (width/two), lesser of the two
U = Velocity of shock wave (M . a)
a = velocity of echo (344 m/s) at MSL at 20°C
M = Mach number of wave given by $\sqrt{(1 + 6p_{so}/7p_a)}$ (2)

4.2. *Side Face Loading*

For side walls, the blast wave is described as a moving triangular wave with the maximum value of overpressure ($p_s + C_d$q) over time t_d [5]. The calculated blast parameters are mentioned in table 3.

Table 3: Blast Load Parameters

S.No.	Parameters	Value

1.	Scaled distance (x)	32.32 m
2.	Ambient air pressure (P_a)	1 kg/cm^2
3.	Peak side on overpressure (P_{so})	1.25 kg/cm^2
4.	Peak reflected overpressure (P_{ro})	3.62 kg/cm^2
5.	Dynamic pressure (q_o)	0.47 kg/cm^2
6.	Scaled time (t_o)	11.35 ms
7.	Duration of equivalent triangular pulse (t_d)	7.7 ms
8.	Front wall loading	3.7 kg/cm^2
9.	Side wall loading	1.1 kg/cm^2

5. Analysis

Blast load function is defined as the time history function in ETABS. Hinges are allocated to columns and beams at a relative distance of 0.1 and 0.9 from either joint. Direct integration analysis based on the non-linear time history is carried out using ETABS software tool.

Blast load is applied as the impact load in the time duration of 7.7 milliseconds on the considered frame as shown in the figure 7. The values of P_{ro} and P_{so} are multiplied with the transverse dimension of beam and column and applied in the units of kN/m. The time history impulse of P_{ro} and P_{so} is shown in figure 5 and figure 6.

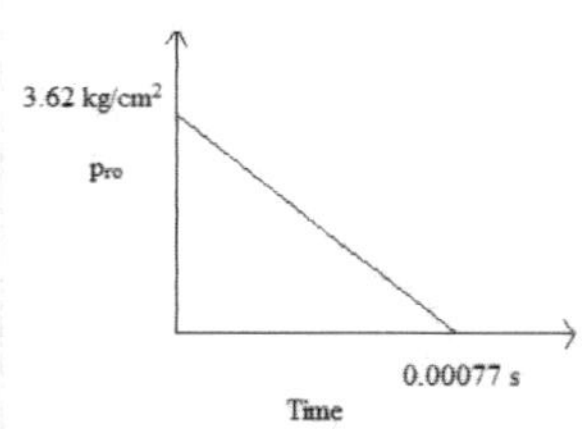

Figure 5: Blast wave on front wall

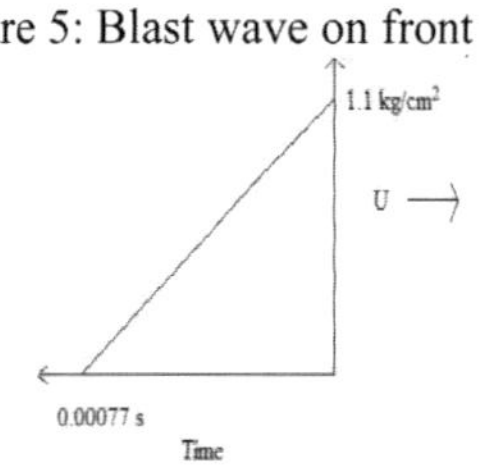

Figure 6: Blast wave on side wall

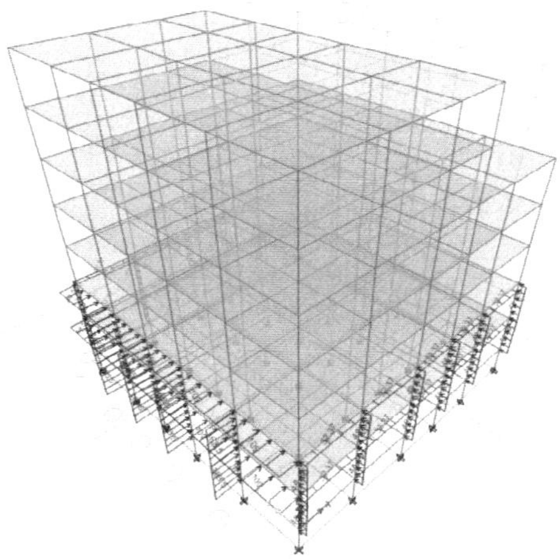

Figure 7: Application of blast load

6. Results

The considered frames suffer deterioration with detriment of beams and columns. The structural elements in the closer vicinity experienced more damage due to high blast pressures. The storey displacement is found to be maximum in Moment Resisting Frame (MRF) and minimum in case of Buckling Restrained Braced Frame (BRBF) as shown in figure 8. The maximum storey displacement in MRF, CBF, EBF and BRBF is 2.2cm, 1.45cm, 1.31cm and 1.1cm respectively. The plastic hinge analysis is also performed and the number of members failed is found to be maximum in MRF and least in BRBF as shown in figure 9(i), figure 9(ii), figure 9(iii) and figure 9(iv).

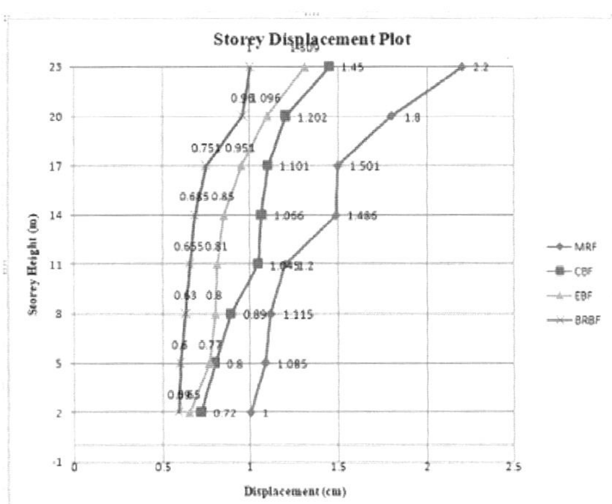

Figure 8: Storey displacement v/s Storey height

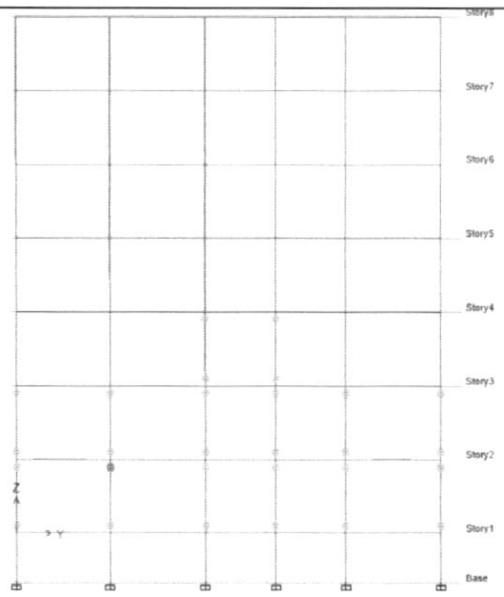

Figure 9 (i) : MRF

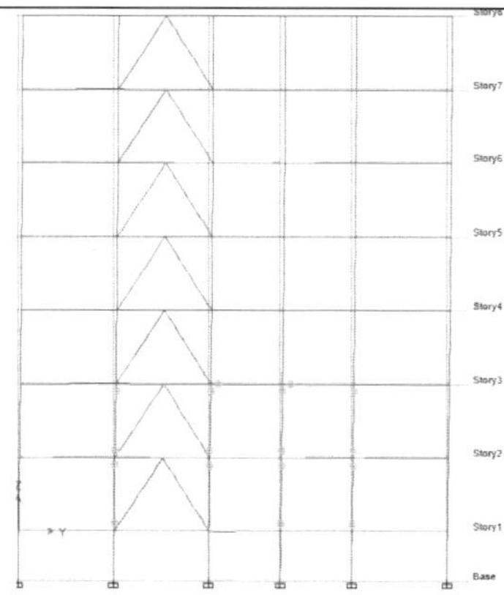

Figure 9 (ii) : MRF

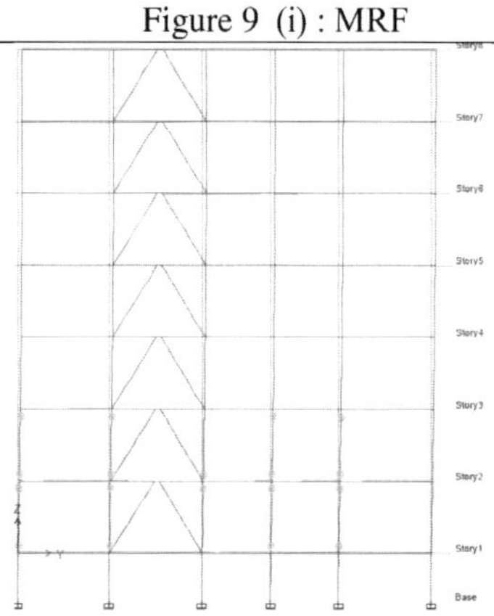

Figure 9 (iii) : EBF

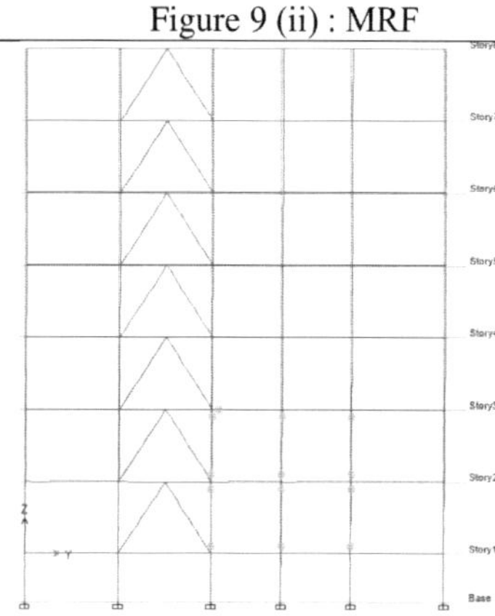

Figure 9 (iv) : BRBF

The plastic hinges are shown in green dots in figure 9(i), 9(ii), 9(iii) and 9(iv). The number of members failed in MRF, CBF, EBF and BRBF is 14, 10, 10 and 7 respectively.

7. Conclusion

Based upon the results obtained from the analysis of four seismically designed frames, it is concluded that:

- The maximum storey displacement in CBF is around 34% less than that in MRF and in case of EBF it is around 40.4% less as compared to MRF.
- The maximum storey displacement in BRBF is around 50% less than that in MRF.
- The maximum storey displacement in BRBF is around 24.1% and 16.03% less than that in CBF and EBF respectively.
- The number of structural members failed in CBF and EBF is approximately 28.6% lesser than that in MRF.
- The number of structural members failed in BRBF is approximately 50% lesser than that in MRF.
- The number of structural members failed in BRBF is approximately 30% lesser than that in CBF and EBF.

Hence the braced frames perform better as compared to MRF under blast loading. The special type of braced frames that is BRBF resisted perform to blast loading to a greater extent to other type of braced systems.

Acknowledgement
I express my gratitude to almighty for his blessings to accomplish such endeavour. I convey my sincere gratitude to my guide Dr. Saraswati Setia, Professor, Civil Engineering Department, NIT Kurukshetra, Haryana for her advice and guidance throughout the study period. Without her, this study would have been a little success.

References
[1] Coffield A and Adeli H 2014 An investigation of the effectiveness of the framing systems in steel structures subjected to blast loading *Journal of civil engineering and management,* 20(**6**): 767–777.
[2] Surendran N and Subha K 2017 Seismic evaluation of building with buckling restrained braces (brb) *International journal of scientific & engineering research,* 8(**11**): 80–86.
[3] UFC 3-340-02 *Structures to resist the effects of accidental explosions*, US Department of Defense (DoD), Washington DC, 2008.
[4] FEMA 2003 *Reference manual to mitigate potential terrorist attacks against buildings*, Standard No. 426, Federal Emergency Management Agency, Washington D.C.
[5] Indian Standard. 1968 *Criteria for blast resistant design of structures for explosions above ground*, Standard. No. 4991, Bureau of Indian Standards, New Delhi.

IOP Conf. Series: Materials Science and Engineering 1004 (2020) 012009 doi:10.1088/1757-899X/1004/1/012009

Shear Mode Damper Testing Using Flake Shape Based Magnetorheological Fluids

D. M. Patel[1*], D. V. Bhatt[2] and R. V. Upadhyay[3]
[1]C S Patel Institute of Technology, Department of Mechanical Engineering, Charotar University of Science and Technology, CHARUSAT-Campus, Changa-388421, India.
[2]Department of Mechanical Engineering, S.V.N.I.T., Surat-395007, Gujarat, India.
[3]P D Patel Institute of Applied Science, Charotar University of Science and Technology, CHARUSAT-Campus, Changa-388421, India
*E-Mail: dipalpatel.me@charusat.ac.in

Abstract: Use of anisotropic iron particles in the magnetorheological (MR) fluid having a high yield stress is a challenge as it increases the viscosity of the fluid in off-state. In this clause, a novel flake shaped iron powder based MR fluid with high yield stress is synthesized and used in shear mode MR damper (SMMD). MR damper design is optimized using fluid properties and Bingham model. Damping performance of newly synthesized MR fluid damper is at par with the available friction based damper used in front loaded washing machine. The unique feature of this design is minimal volume of the fluid (1.5 ml) required to achieve damping force of 50N. Effect of different volume percentage of particles in MR fluid is evaluated in terms of damping force. Effect of testing parameters like displacement and frequency of excitation on damping force was evaluated. Effect of settling of particle on damper performance is discussed in this work. It shows that the flakes shaped-based MR fluid show better stability against gravity. The smaller quantity of present flake shaped-based MR fluid will reduce the cost of shear mode damper.

1. Introduction

By using passive damper between structure and delicate equipment, it is possible to take in either shock (high damping) or vibration (low damping) during normal operation. Therefore, research was concentrated on developing active dampers for vibration isolation [1]. Semi-active damper utilizing Magnetorheological (MR) fluids also can be used for this requirement. The MR fluids are combinations of carrier oil, iron particles (micron-Sized). It's rheological properties are tune due to magnetic particles in presence of magnetic field [2,3]. MR effects (rheological properties) in the MR fluid has found numerous applications like MR shock absorbers [4], MR mounts [5], clutches, brakes [5,6], recoil system [7-9] etc.

In all these applications, the maximum effect is limited by the saturation magnetization of the dispersed magnetic particles. The performance of MR fluid in applications is judged by three parameters called figure of merits [10]. These are; (i) $F_1 = \tau_2/\eta$, where τ is yield stress and η is the viscosity of MR fluid at zero magnetic field, maximizing this figure of merit, helps in minimizing the device size and electrical power consumption. (ii) $F_2 = \tau_2/\eta\rho$, here ρ is the density of MR fluid. This is an important term, when one deals with weight-sensitive applications and (iii) $F_3 = \tau/BH$, here B and H are magnetic flux density and intensity of the fluid. Minimization of this figure of merit will help to reduce current requirement, and it is usually required for high bandwidth applications. Then all these virtues are to be optimized for the MR fluid properties. While the suitability of an MR fluid for a particular use depends on factors such as stability, durability and temperature range, therefore, it becomes very significant to design new MR fluids, which can enhance the MR effect.

In earlier research, much attention has been focused on the use of spherical carbonyl iron particles as a dispersion material because of its high saturation magnetization value. More frequently than not, good MR fluids should be stable against settling and should have high magnetic strength. The major focus on MR fluid research has been to enhance the MR effect, to improve stability against sedimentation and re-dispensability of MR fluids [10-19]. In summation, many efforts have been reported by adding nanoparticles and viscoelastic media to improve stability and re-dispersion. According to recent research, MR fluid based on micro- and nano-fibres have shown better sedimentation stability and higher yield stress comparable to spherical particles for a given volume fraction and magnetic field intensity [20-23].

Furthermore, our recent works, reported the suitability of flake shaped particles for MR fluid synthesis and it's an application in device [22-23]. The study proposes that the MRF exhibit an improvement in τ at low field with <0.1% per day sedimentation rate. In addition, flake shaped iron particles required low magnetic field to reach the saturation, compared spherical iron particle-based MRF.Having a high value of saturation magnetization of spherical iron-carbonyl powder, in all the MR based applications, the MR fluids are synthesized using these powders. Use of non-spherical shape iron powders in developing MR based applications are very uncommon. The major difficulty of using the anisotropic particles is to get a high value of the first figure of merit F1. This is imputable to the growth in the viscosity as well as a limitation of the higher volume fraction of particle loading. Few works, reports on applications of flakes-shaped (plate-like) iron particles dispersed MR fluid for damping applications [22-24]. At all this reported work, the damping force is depressed. As a result, the required volume in the damper becomes high to get the same damping force values, which can be obtained using spherical iron particles.

The primary focal point of current work is first to improve the damping performance using flake shaped particle-based MR fluid. To achieve the first aim, we have modified the synthesis conditions of flake-shaped MR fluid and studied its MR properties. In the present paper, we concentrate on shear mode MR damper (SMMD), which can be used as washing machine damper. In the present condition, in front-loaded washing machine friction damper is used and test result of this damper is as shown in figure 1. Friction damper is tested at 1Hz frequency with 5mm of stroke length this damper can generate 40N of force which is not tuneable with the different operating speed of the washing machine. Therefore, in the present work design of SMMD based on the above result is carried out and tested. Present work is also identifying the influence of volume % of particles on MR damper to enhance the performance of damper as well as effect of different test parameters on performance of damper is discussed in details. In next section theory related to SMMD is discussed and MR fluid properties followed by result and discussion.

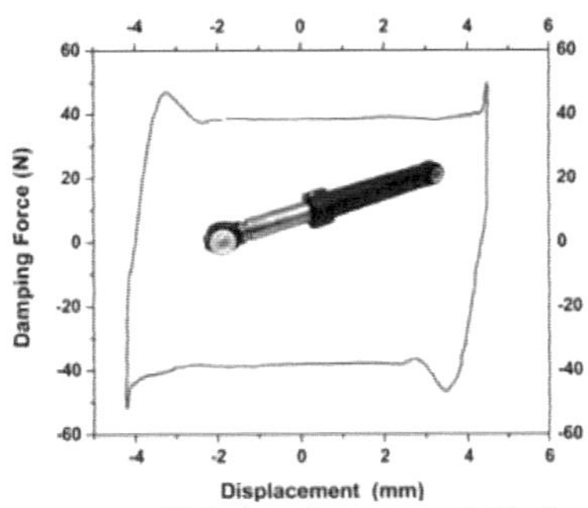

Figure 1: Force- displacement curve of friction damper at 1 Hz frequency and 5 mm displacement.

2. Theory

Bingham model and parallel plate concept (Figure 2) are used to represent damping force [23-25],

$$F = (\tau_y(B) \times L \times b) + (L \times b \times \mu \frac{v}{t}) \tag{1}$$

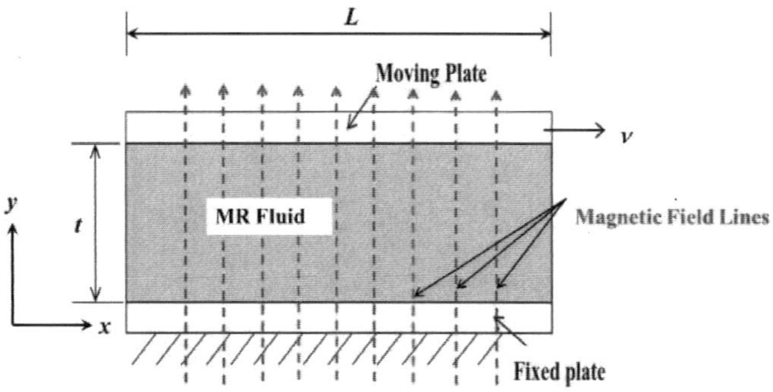

Figure 2: Parallel plate configuration.

Where $\tau_y(B)$ is shear stress of MR fluid at a given B. μ is viscosity of MR fluid. v is velocity of moving plate. t is the gap between the plates, L=length of plate. b= breadth of plate.

By using equation 1 design of SMMD is carried out. The design consists of electromagnetic coil and piston with outer casing (figure 3). In between piston and electromagnetic coil MR fluid is placed. Gap between these two components is MR fluid gap (g), which kept 1mm. In this configuration comparing with parallel plate thus, piston is acts as moving plate and velocity is represent as v_p and electromagnetic coil acts at stationary plate. By substitute the width (b) with $2\pi D_P$ and length (L) with 'L_e'. By considering this equation, 1 can be re-written as, so damping force.

$$F = (\tau_y(B) \times L_e \times \pi D_p) + (L_e \times \pi D_p \times \mu \frac{v_p}{g}) \tag{2}$$

Where, D_p is diameter of piston; First portion of the equation 2 shows the force is the depends on 'τ_y' and it depends on 'B' (magnetic field). Second portion of equation 2 shows force is not a function of 'B'. It is also known as $F_{off\text{-}state}$. Equation 2 also known as $F_{on\text{-}state}$.

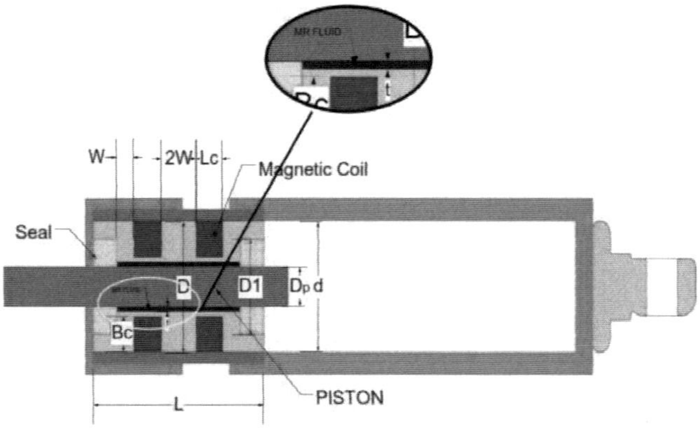

Figure 3: Cross section view of SMMD.

From the above mention, equation 2 design of MR damper is carried out and optimized as per need of damping force 40N which is generated in presently available washing machine damper. The finalized dimensions are presented in Table 1.

Table 1: Dimensions of the SMMD.

Parameter	Symbol	Dimension in mm
Outer diameter of cylinder	D	50
Inner diameter of cylinder	D_i	40
Effective length of piston		36
Diameter of piston	D_p	5
Outer pole length	W	5

Coil height	L_c	8
Coil width	B_c	8
Length of magnetic coil	L	50

For any kind of MR damper magnetic flux produce due to external current in MR fluid gap is very important. Therefore, Using '*COMSOL Multiphysics*' software magnetic field simulation is carried out. For the simulation, the material used for the electromagnetic coil is having high magnetic permeability. The distribution of magnetic field (MF) is as shown in figure 4. Magnetic flux dissemination in MR fluid gap is represent with line graph at different current is as shown in figure 5. Figure 5 shows the distribution of MF along the MRF gap is nearly constant. Maximum flux is at piston and at outer cylinder, which is nearly 0.6T. The value of flux at MR fluid is 0.21T.

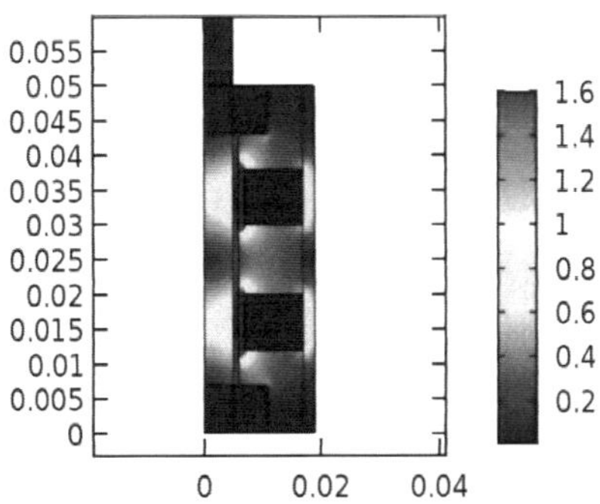

Figure 4: Magnetic flux density (T) distribution in SMMD.

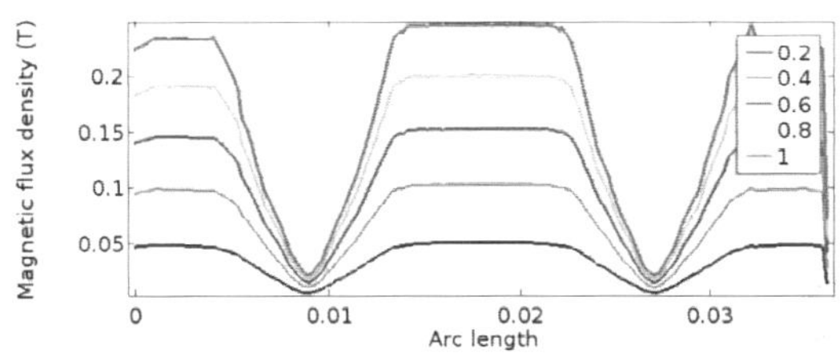

Figure 5: Magnetic field B(T) in MR fluid gap at 0.2A to 1A

MR fluid properties use in this simulation is as shown in Figure 6. In this paper two different volume percentage, based MR fluid is used for the testing so behaviour of shear stress (yield stress) with different MF is as shown in Figure 6. Table 2 shows the different properties of MRF18 and MRF26. Based on the dimensions in table -1 SMMD is manufacture. All the parts and assembly of SMMD are as shown in Figure 7.

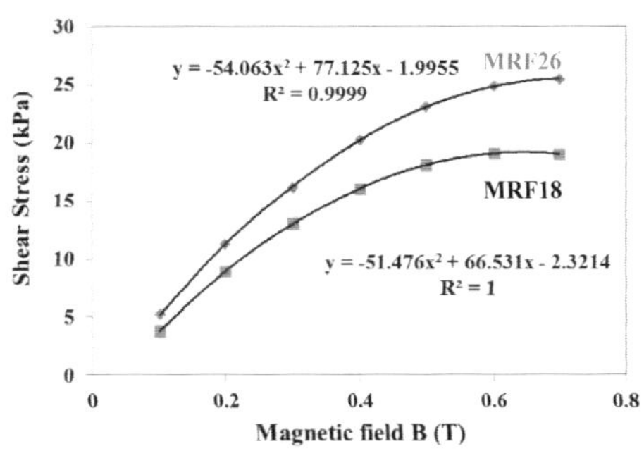

Figure 6: Behaviour of shear stress with magnetic field.

Table 2: Properties of MRF18 and MRF26.

Properties	Values	
	MRF18	**MRF26**
Maximum yield stress (kPa)	18±01	26±01
viscosity (mPa. s) @ 40°C	68±01	98±01
Volume percentage (%)	15	22
Magnetic permeability, relative @low Field	3.5	4

Flash point (°C)	>150	>150

3. Results and Discussions

3.1 Testing of Damper

Testing of SMMD is performed using damper testing machine (AGMPL-15027) (figure 8). Sinusoidal excitation of fixed frequency and amplitude (stroke length of damper or displacement) is generated using the hydraulic drive. Load cell is measured the force. Linear displacement is sensed using LVDT. DC power supply is used for energies the electromagnetic coil to produce the MF.

For testing MR fluid filled in damper is 1.5 ml. After that, the damper was set along the test rig and initially, three cycles were recorded. The average of the three cycles has been counted for final plot of force versus displacement curve.

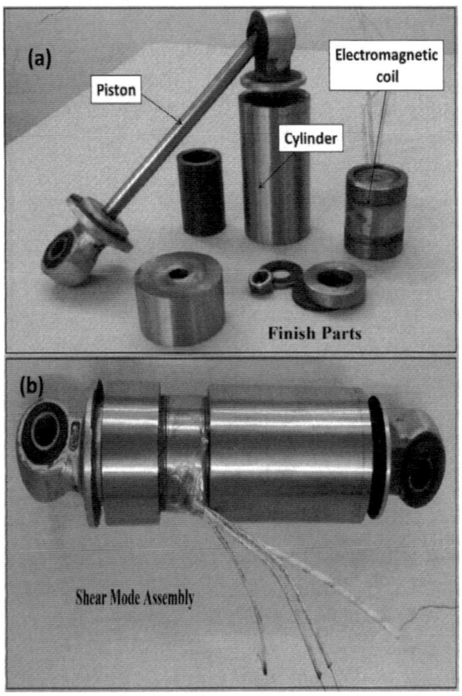

Figure 7: Manufacture components and Assembly of SMMD.

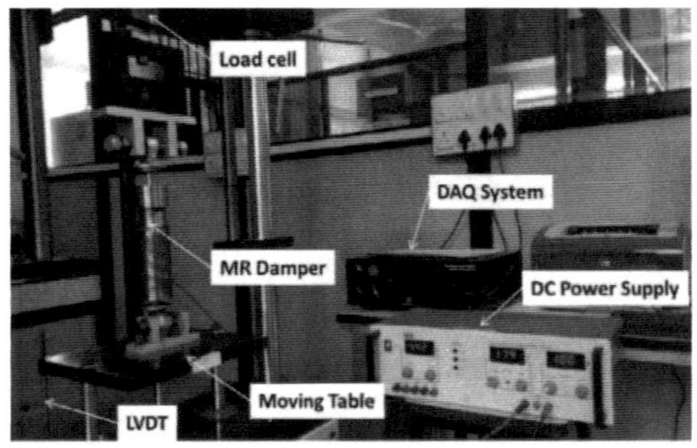

Figure 8. Test rig for damper test.

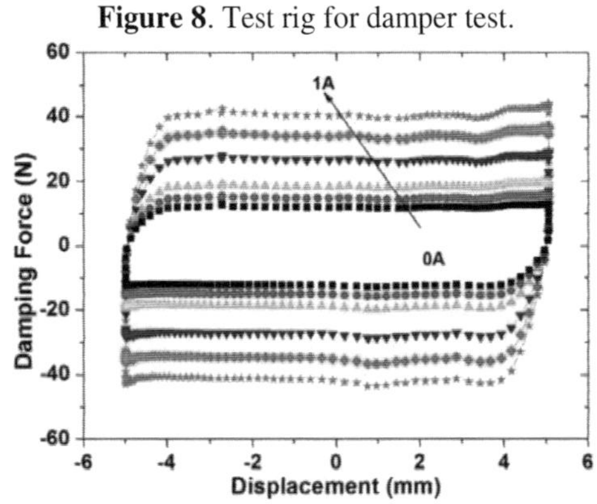

Figure 9: Force-displacement plot for SMMD MRF18.

3.2 Effect of Applied Current

First, MR damper tested at 0A current. In figure 9, black scatter point shows maximum force is 12N in rebound, and minimum is -11N in at B=0 (No current is supplied to the coil). This condition is considered as off-state condition where B=0 ($F_{off\text{-}state}$). This is only due to the friction between the piston and rubber oil seal. As the current increases MF in the MR fluid gap increases, which result in to rise in yield stress thus force rises. In the presence of MF, particles are magnetize and orient in the direction of MF. Which help to formulate the chain of particles, which generates the column like structure. In the presence of column like structure, fluid becomes semi-solid and provide resistance to motion of the piston. The strength of the columnar structure depends upon the MF strength. As current increases, the MF strength increases so the strength of columnar structure increases, which results in higher damping force. Damping force is 42N and -43N at 1A current for MRF18. For this test frequency of excitation is kept 1Hz.

Figure 10 shows the comparison between experimental data with theoretical model from equation 2 for 0A to 1A current. Good similarity observed between both data. It validates the theory. As current increases due to MR effect, force increases.

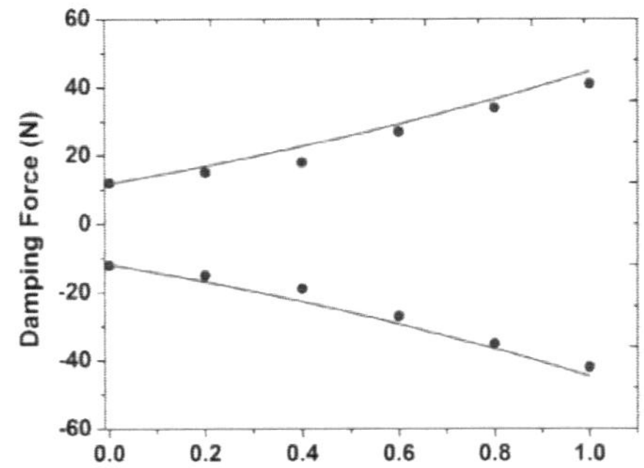

Figure 10: Damping force -current with theoretical comparison for MRF18.

3.3 Effect of Volume Percentage of Particles

MR damper performance is depended upon many parameters one of the parameters is the volume percentage of iron particles added in the carrier fluid. As the volume percentage increases the yield stress is higher, due to strength of the columnar structure is more; thus, damping force increases. To investigate this MR fluid with 22% volume is synthesised and the properties of this fluid is discussed in table -2. Volume of MRF18 is 15%. With higher volume percentage, MRF26 generates the higher damping force compare to the MRF 1 at same current.

The test result of MRF26 with same testing parameters is as shown in figure 11 and 12. MRF26 force is 53N and -52N at 1A current (Figure. 11). For this frequency of oscillation kept 1Hz. For both the fluid, force at no current is same. Figure 13 shows the deviation in force with current for MRF18 and MRF26. Polynomial equation from the test results is represented with equation 3 for MRF26 and equation 4 for MRF18.

$$F_d=20.98*(I)^2+17.30*(I)+12.72 \quad (3)$$

$$F_d=18.30*(I)^2+12.54*(I)+11.67 \quad (4)$$

Where, *I* is current.

3.4 Effect of Different testing parameters

To investigate the effect of different testing parameters like stroke length (displacement) and frequency of sinusoidal excitation. Figure 14 shows the force at a fixed frequency of 1Hz and 1A current for different displacement.

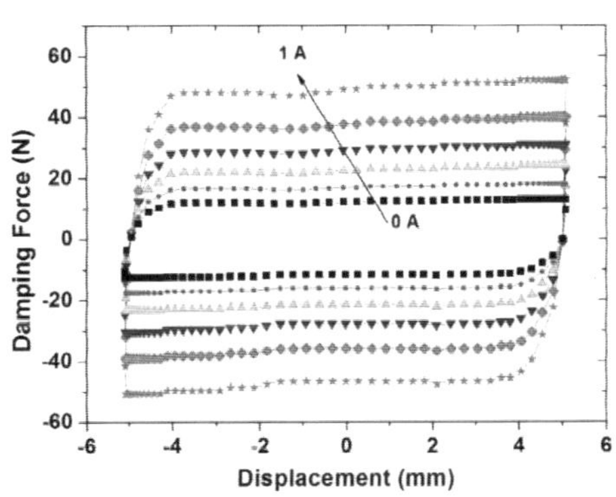

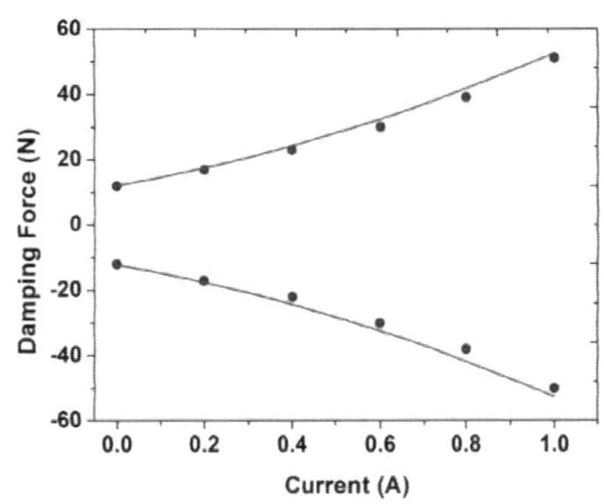

Figure 11: Force-displacement for SMMD for MRF26.

Figure 12: Damping force -current with theoretical comparison for MRF26.

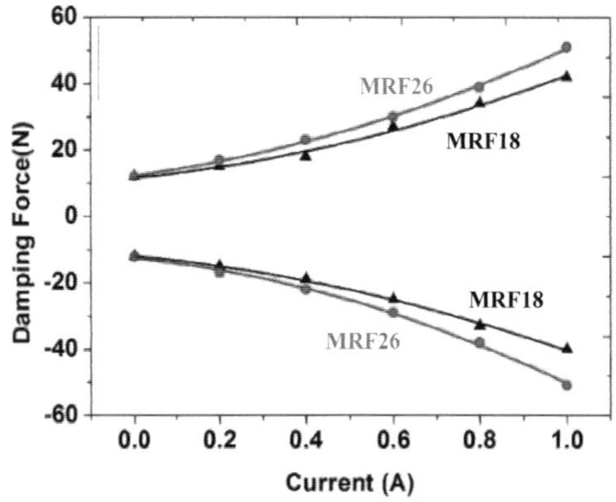

Figure 13: Damping force -current for MRF18 and MRF26.

Change in damping force with displacement is very low so one can say SMMD damping force does not depend upon the displacement. Deviation in damping force for displacement of 5mm to 16mm is nearly ±2N, which is within the range of error bar of instrument. At 5mm displacement damping force is 50N for MRF26 and 42N MRF18 and at 16mm damping force is 53N and 43N for MRF26 and MRF18 respectively. In figure 14 linear fitted equation shows the maximum deviation in damping force with per millimetre of displacement for MRF18 is 0.17 (slope of the line) and for MRF26 is 0.13.

Another important testing parameter is frequency of excitation. To investigate this kept the current and displacement is fixed (1A and 10mm). Figure 15 shows the change in force at various frequency is linear. Change in damping force with frequency is also lower but slightly higher compared to displacement. This deviation is range from 50N to 54N from the range of 0.2Hz to 1Hz for MRF26 and 38N to 42N for MRF18. Deviation in damping force with frequency is ±3.5 for MRF18 and 2.5 for MRF26.

3.5 Effect of time

Use of flake shaped particles also influence on sedimentation rate. To investigate this test is performance with the damper at different time interval. For this MR damper is tested at 0A current and 1A current at fixed frequency of 1Hz and 10mm displacement for different three-time interval. First damper is filled with MR fluid and tested, which mention as 0hr test. Than retain, the damper in stable condition and same test is conducted after 24hr and collect the force displacement plot. Same test is performed after 48 hr and 72hr.

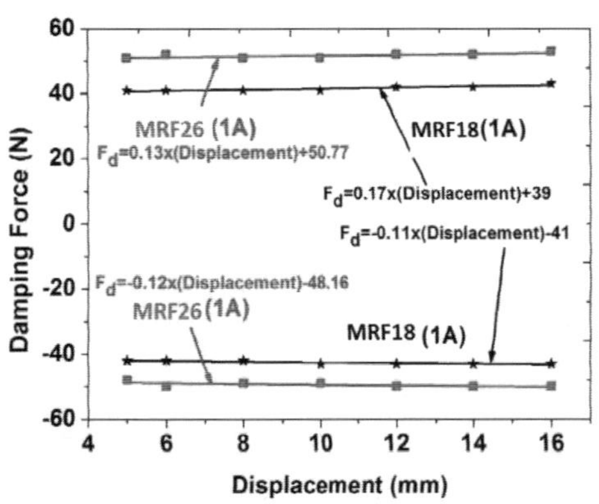

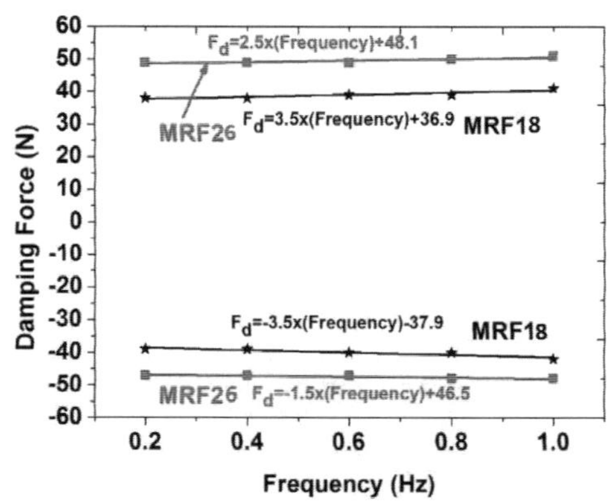

Figure 14: Damping force -displacement at frequency of 1Hz and 1A current for MRF18 and MRF26.

Figure 15: Damping -frequency at 1A current and 5 mm stroke length for MRF18 and MRF26.

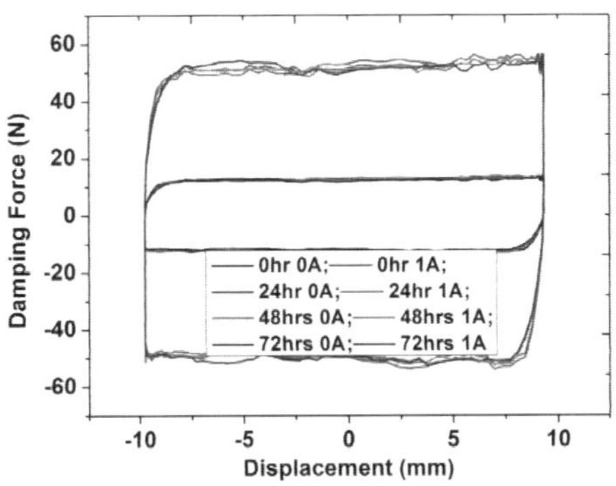

Figure 16: Damping force-displacement curve after a different time interval.

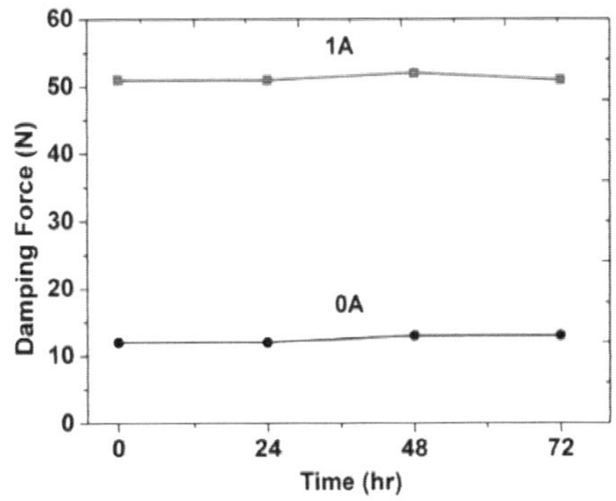

Figure 17: Damping force at different waiting period time at 0A and 1A current.

Figure 16 shows the variation in damping force of SMMD at 0hr, 24hrs,48hrs, and 72hrs for 0A and 1A current. All the test results are almost similar. It shows better stability of MRF26 in damper application. Figure 17 shows the deviation of damping force at different time interval and it is nearly straight horizontal line; thus with time, the effect of settling of the particles is very less. Damping force remains unchanged at 0A and 1A current after different time interval.

4. Conclusions

In this work design of SMMD is carried out based on the MR fluids which is synthesised using different volume percentage of flaked shaped iron particles. For design of SMMD friction damper of front-loaded washing machine is considered as reference. Based on the optimized design, the damper is manufactured and tested on damper testing machine. Test result shows that with increases in applied current damping force of SMMD increases. Variation of damping force also depends on the volume percentage of particles and applied current. From the results, it can be concluded that using MRF18 in SMMD, on-state force is three times higher compare to off-state condition. Similarly, for MRF26 ratio of on state to off-sate force is nearly four time higher, which conclude that with higher volume percentage gives higher damping force due to higher yield stress. Stroke length and frequency effect on force is very small. From that, one can say for constant damping force at different stroke and frequency SMMD provides best solution. Flake shaped based fluid also shows the better stability in terms of damping performance at different time interval. Last, we can conclude that, SMMD fluid volume required for the damping is nearly 1.5ml. Which can reduce the cost of damper and design SMMD can be used in washing machine application.

Acknowledgement

Authors are thankful to Charotar University of Science and Technology (CHARUSAT), Changa and Industrial metal powder India Pvt. ltd, Pune for financial support for experimental facilities.

References

[1]Park B J, Fang F F and Choi H J 2010 Magnetorheology: materials and application *Soft Matter* **6,** 5246-5253.
[2]Wereley N 2013 *Magnetorheology: advances and applications* **6,** (Royal Society of Chemistry).
[3]Kim H C, Oh J S and Choi S B 2014 The field-dependent shock profiles of a magnetorhelogical damper due to high impact: an experimental investigation *Smart Mater. Struct.* **24,** 025008.
[4]Chen P, Bai X X, Qian L J and Choi S B 2016 A magneto-rheological fluid mount featuring squeeze mode: analysis and testing *Smart Mater. Struct.* **25,** 055002.
[5]Nguyen Q H and Choi S B 2012 Optimal design of a novel hybrid MR brake for motorcycles considering axial and radial magnetic flux *Smart Mater. Struct.* **21,** 055003.
[6]Sarkar C and Hirani H 2013 Theoretical and experimental studies on a magnetorheological brake operating under compression plus shear mode *Smart Mater. Struct.* **22,** 115032.
[7]Singh H J and Wereley N M 2014 Optimal control of gun recoil in direct fire using magnetorheological absorbers *Smart Mater. Struct.* **23,** 055009.
[8]Wereley N M, Choi Y T and Singh H J 2011 Adaptive energy absorbers for drop-induced shock mitigation *J. Intel. Mat. Syst. Str.* **22,** 515-519.
[9]Carlson J D, Catanzarite D M and St Clair K A 1996 Commercial magneto-rheological fluid devices *Int. J. Mod. Phys. B* **10,** 2857-2865.
[10]Jolly M R, Bender J W and Carlson J D 1998 Properties and applications of commercial magnetorheological fluids *Smart Structures and Materials 1998: Passive Damping and Isolation* **3327** 262-276.
[11]De Vicente J, López-López M T, González-Caballero F and Durán J D G 2003 Rheological study of the stabilization of magnetizable colloidal suspensions by addition of silica nanoparticles *J. Rheol.* **47,** 1093-1109.
[12]Pu H T, Jiang F J, Yang Z, Yan B and Liao X 2006 Effects of polyvinylpyrrolidone and carbon nanotubes on magnetorheological properties of iron-based magnetorheological fluids *J. Appl. Polym. Sci.* **102,** 1653-1657.
[13]Wereley N M, Chaudhuri A, Yoo J H, John S, Kotha S, Suggs A, Radhakrishnan R, Love B J and Sudarshan T S 2006 Bidisperse magnetorheological fluids using Fe particles at nanometer and micron scale *J. Intel. Mat. Syst. Str.* **17,** 393-401.
[14]López-López M T, Zugaldia A, González-Caballero F and Durán J D G 2006 Sedimentation and redispersion phenomena in iron-based magnetorheological fluids *J. Rheol.* **50** 543-560.
[15]López-López M T, Vicente J De, Bossis G, González-Caballero F and Durán J D G 2005 Preparation of stable magnetorheological fluids based on extremely bimodal iron–magnetite suspensions *J. Mater. Res.* **20,** 874-881.
[16]López-López M T, Kuzhir P and Bossis G 2009 Magnetorheology of fiber suspensions. I. Experimental *J. Rheol.* **53,** 115-126.
[17]Ngatu G T and Wereley N M 2007 Viscometric and sedimentation characterization of bidisperse magnetorheological fluids *IEEE Trans. Magn.* **43,** 2474-2476.
[18]Ngatu G T, Wereley N M, Karli J O and Bell R C 2008 Dimorphic magnetorheological fluids: exploiting partial substitution of microspheres by nanowires *Smart Mater. Struct.* **17,** 045022.
[19]Viota J L, González-Caballero F, Duran J D G, and Delgado A V 2007 Study of the colloidal stability of concentrated bimodal magnetic fluids *J. Colloid Interface Sci.* **309,** 135-139.
[20]Bell R C, Karli J O, Vavreck A N, Zimmerman D T, Ngatu G T and Wereley N M 2008 Magnetorheology of submicron diameter iron microwires dispersed in silicone oil *Smart Mater. Struct.* **17,** 015028.

[21]Upadhyay R V, Laherisheth Z and Shah K 2013 Rheological properties of soft magnetic flake shaped iron particle based magnetorheological fluid in dynamic mode *Smart Mater. Struct.* **23,** 015002.
[22]Shah K, Oh J S, Choi S B and Upadhyay R V 2013 Plate-like iron particles based bidisperse magnetorheological fluid *J. Appl. Phys.* **114,** 213904.
[23]Shah K, Seong M S, Upadhyay R V and Choi S B 2013 A low sedimentation magnetorheological fluid based on plate-like iron particles, and verification using a damper test *Smart Mater. Struct.* **23,** 027001.
[24]Patel D M and Upadhyay R V 2020 Testing and evaluation of linear shear mode magnetorheological (MR) damper based on rheological properties of MR fluid. *J. **Test** Eval* **48**.
[25]Wereley N M, Cho J U, Choi Y T and Choi S B 2007 Magnetorheological dampers in shear mode *Smart Mater. Struct.* **17,** 015022.

Artificial Intelligence for Compressive Strength Prediction of Concrete

S. J. Goutham[1*] and V. P. Singh[2]

[1] National Institute of Technology Kurukshetra, Haryana, India.
[2] National Institute of Technology Kurukshetra, Haryana, India.
*Email: gsjsai@gmail.com

Abstract. Structural health monitoring is an indispensable procedure that is to be carried out to evaluate the serviceability of existing structures. Non-destructive testing methods are attaining increasing popularity for the assessment of concrete strength due to the ease of operation and reliability of the results. The application of machine learning in the field of engineering has increased rampantly. In this study, the compressive strength of concrete has been forecasted using support vector regression which is a machine learning technique. Artificial intelligence is nothing but the potential to impersonate human intelligence. The advantage of artificial intelligence over human intelligence is the absence of human errors that can arise due to various factors which might decrease the accuracy of results. Rebound hammer (RBH) , Windsor probe penetration (WPP) and ultra-sonic pulse velocity(USV) are the non-destructive testing that has been employed to assess the concrete compressive strength. The use of multiple non-destructive testing methods than single testing methods has improved the accuracy of prediction model. Further the results of different combination models were compared using coefficient of determination which is a commonly used statistical parameter for prediction accuracy comparison. The prediction model is in accordance with the experimental results obtained. And the accuracy of prediction indicates that support vector regression can be successfully used for predicting the compressive strength of concrete.

1. Introduction

Engineering is occluded with hurdles that cannot be easily solved using ordinary typical computational methods. Artificial intelligence is the solution to such problems. It is a computational method that strives to replicate human comprehension capacity and creates outputs that challenges traditional normal computational techniques. Cognitive computing is another bough of artificial intelligence which was influenced by human mind's potential. The problem solving nature of cognitive system and human minds are similar viz. thinking and reasoning. The main attributes of such systems are their capacity to decode large data, training dynamically and learning adaptively which helps in establishing a consistent pattern. The use of artificial intelligence in the field of research has grown rapidly in the past decade.

The compressive strength of concrete at 28 days is a decisive parameter in RCC. The conventional procedure to procure the concrete mix design is empirical and performance based. Unambiguous evaluation of concrete strength at 28 days is advisable for better control of quality. But when it comes to the strength prediction of concrete in actual structures, the process is not so simple due to the irregular environmental conditions and other variations that can transpire during mixing, placing, compaction and finishing. Prediction of concrete strength is a multifaceted problem. Hence in-order to contemplate all the uncertainties involved, it is desirable to go for a non-linear mathematical model. The use of machine learning for concrete compressive strength prediction has been acclaimed widely due to the uncontrollable variations involved in designing of concrete mix. One of the popular machine learning algorithms for concrete strength prediction is artificial neural network (ANN). Though ANNs have given satisfactory values of prediction, it has the tendency of data overfitting for datasets that are smaller. Such an overfitting can lead to powerful performance of the training set but a substandard performance in generalization on new dataset [1]. Support vector regression (SVR) is yet another machine learning tool that has been extensively used for strength prediction. The shortcomings in prediction using ANN has been resolved in SVR. Existing structures have to be evaluated for their structural performance in order to assess its seismic performance and for its retrofitting after a damage or failure. For this evaluation, the mechanical properties of concrete have to be assessed. Estimation through coring has its limitations due to technical difficulty and expense. Non-destructive testing provides a feasibility alternative for assessment of in-situ strength of concrete. Relationship which is empirical in nature is identified between the strength assessed on the concrete core and the NDT result. For quality in estimation of strength, the empirical relationship identified must be error free and valid for any input dataset. Presence of uncontrollable parameters, type of NDT, quality of measurement is few of the many parameters which influences the reliability of estimation [2]. In this study, concrete cubes of various strength were casted and three NDT viz. RBH, WPP and USV were employed on these cubes. The results of NDT were used to generate a prediction model using SVR in python which accurately predict the compressive strength of concrete using these NDT results. NDT results were used individually as well as in combination to evaluate the performance of these NDT techniques. Coefficient of determination (R^2), root mean squared error (RMSE), mean absolute percentage error (MAPE) and mean absolute deviation (MAD) were the statistical parameters identified to derive a comparison between various models.

2. Literature Review

Advanced data mining techniques have a huge scope in the field of engineering for problem solving. Important data mining techniques include prediction model (machine learning implementing supervised learning), association and clustering (unsupervised learning) and so on. Problems associated with civil engineering are diverse and complex. It can be attributed to extremely variable factors. In order to address these issues, machine learning algorithms have been developed that can provide satisfactory solution to these complex problems which will be difficult otherwise to solve using conventional techniques. The process of fine tuning the parameters associated with the algorithm is a salient process. A prediction model is nothing but an algorithm that establishes a distinct relationship with the input and output parameters and which can successfully be used for the prediction for further data. For instance, forecasting the properties of any material in a crucial function in material science. Due to the high learning capabilities of support vector regression, it has found a prime place in the field of data prediction [3]. Neural networks and SVR models has been generated to predict the strength of no-slump concrete and although both the models gave sufficient prediction accuracy, SVR model is preferred over ANN due to its various superior features [4]. Research has also verified the use of SVR for computing the strength of high performance concrete. Experimental results have yielded low prediction errors for the same. Such a prediction model can be used efficiently to identify concrete mixtures of substandard strength[5]. NDT methods have evolved as a response to the detection of structural damage and its prevention. Development in sensors and new materials and reduction in size of devices are improving the scope of developing new NDT methods. Assessment of material integrity using NDT is attaining popularity[6]. From literature survey, it has been found out that fuzzy logic can also be used for generating prediction models to forecast the compressive strength of cement mortar. When the fuzzy logic model was compared to that of ANN, it was discovered that the former model could predict data with a lower errors and also the model generated was more user friendly [7]. Machine learning is just not a knowledge generation tool. It can also be efficiently used for analysis of data. It is very crucial to perceive the exact nature of problem. It can be better studied by analysing previous applications. A comprehensive understanding of machine learning process can be used for incorporating the features of the problem into the techniques of machine learning. To achieve this there is constant need to update with the recent trends and findings in the field of machine learning. Machine learning techniques can be enhanced by carrying out comparison studies. Evaluation of any kind involving single or multiple techniques of machine learning which is qualitative or quantitative requires special attention in each step of interpretation, design and analysis. Without this, the output can be faulty and might be influenced by random variables [8].

The application of machine learning is structural health monitoring is highly beneficial and constructive. Health monitoring of structures involves collection of data through devices, uprooting the damage prone characteristics and interpreting the extracted data. The studies in the area of monitoring of structures can be broadly seen as data and model drive. The later adopts a numerical model and the former requires the functioning of machine learning and pattern recognition. It is interesting to note that, data driven approach is more efficient and useful if data involved is huge, the physical traits of the structure is unspecified and difficult to model and there is a need to reduce the effort in computation. In the field of structural health monitoring, the functions of machine learning includes creation of database from existing data , understanding the parameters of the model and then solely focussing on prediction of data. Within the scope of machine learning, in terms of robustness and effectiveness, support vector regression, artificial neural network and principal component analysis are few of the techniques which have attained popularity for identification of structural damage. The properties of concrete is known to have highly complex non-linear relationships between its constituents [9]. Hence there is a need to develop a reliable model for better understanding of the characteristics of the material in such a way that it saves both time and cost [10]. Machine learning algorithms could successfully model the traits of concrete which is self- compacting. Support vector regression has also been used successfully for forecasting the split tensile strength of concrete. RBF and polynomial kernel functions were used for the accurate prediction of split tensile strength. The model generated is very efficient because it can execute regression of non-linear nature for data set of higher dimensions [11]. SVM based models have also been generated to establish the relationship between results obtained from NDT and the strength of concrete. The service life assessment and durability can be predicted using machine learning technique with high level of accuracy. The complicated relationship between various parameters can be learned by machine learning algorithms and forecast its remaining service life with a high degree of reliability [12].

3. Support Vector Regression in Python

Support vector machines were first discovered by Vapnik and Guyon. Along with the strong basis in statistical learning theory, SVM exhibited great accomplishment in various fields. Along with other soft computing methodologies, SVM has also established itself as a most sought after machine learning technique. It falls under the category of controlled learning technique. Kernel functions which are no-linear in nature is used for classification problem for transforming input data to data of higher dimension. Compared to neural network, the development path of SVM is straight opposite. Though initially SVMs were not widely popular despite having a powerful theoretical background, later it gained momentum and proved itself better than other machine learning techniques [13]. Support vector regression (SVR) which is used for regression problems is a variant of support vector machines that is mainly used for classification problems. The model generated in SVR is controlled by a dataset subset and is termed as support vectors. These support vectors generates errors greater than threshold value during training of SVR model. Support vector regression is in-demand technique that has produced reliable results in the field of data prediction.

The algorithms in SVR employ functions that are mathematical in nature which is referred to as kernels. The purpose of such kernel functions is to absorb the provided data as input and mutate into the desired form. Different algorithms use divergent kernel functions. Linear, radial basis function (RBF), polynomial, sigmoid and non-linear are some of the un-built kernel functions in support vector regression. Among the various kernel functions available, RBF kernel is the most sought after kernel due to the better output it delivers and the overall reduction in RMSE. For the prediction developed, RBF kernel function itself has been used. It was decided on the basis of trial and error. The major objective of adaptive regression methods like support vector regression is to establish underlying structure in the dataset. For a defined set of data for training, the objective is to attain a function that has the highest variation from the absolute target for all the data falling under training and simultaneously have the maximum flat which means the errors are neglected as long as they are within the specified limits. Flat refers to LIBSVM is a module used for SVR. Its standard use involves training a dataset to obtain a model that has successfully identified the distinct complex relationship between input and output data. The next step involves predicting the testing dataset. Various extensions of LIBSVM is also available. LIBSVM helps in achieving the target of regression, classification and distribution [14]. SVR can be implemented in various softwares. In the present study, it has been executed in python programming language. Python has established itself as one of the most favoured language for scientific computing. Its popularity can be attributed to its interactive user interface and the vast scientific libraries it possess. Scikit-learn utilize this computing friendly interface to contribute various machine learning algorithms. Hence provides a support for analysing the data by amateurs in the field of machine learning and artificial intelligence. This facilitates the use of artificial intelligence in various branches of engineering. Due to its liberal license policy, python is available as open source software package. Both supervised and unsupervised learning algorithms are provided under scikit-learn [15]. Python's syntax is sophisticated yet simple lets the programmer write in a procedural manner. The interpreter of python is highly interactive which enables live development of code and its simultaneous experimentation. This reduces the time required for code development and also increases the productivity of the programmer. The library modules available in python are plenty. In addition to the inbuilt library, we can download other library functions also and expand the library collection as per our need. The popularity of python as a computational tool can be attributed to its simple and easily understandable syntax. Blocks of code identified through indentation, looping structures that are easy to comprehend are few examples of python's uncomplicated syntax. Apart from the lucid syntax of python, python promotes codes that are easily maintainable by dividing the codes into modules [16]. While training a model in python, two parameters are considered viz. gamma and *C*. The proper choice of gamma and *C* is crucial for desirable performance of SVR. For different combinations of NDT, different values of gamma and *C* have been used. It has been arrived on the basis of trial and error method. Table 1 lists out the combination of *C*, gamma and kernel function for different NDT combination.

Table 1: Parameters of SVR algorithm

NDT Technique	Kernel	C	Gamma
USPV	rbf	9800	0.23
PP	rbf	9800	0.23
RH	rbf	17000	0.067
USPV-PP	rbf	930	0.00032
PP-RH	rbf	930	0.00032
RH-USPV	rbf	930	0.00032
USPV-PP-RH	rbf	1320	0.0011

4. Experimental Work

350 concrete cubes were casted as per the recommendation of IS 10262:2009. Moulds of size 15x15x15 cm were used. Major steps involved in the cube preparation was sieving of coarse and fine aggregates, uniform mixing of the raw materials, placing of concrete, compaction and curing. The fractions of raw materials were varied to achieve samples of various strengths. It was done in order to have a dataset possessing wide range of concrete strength. The minimum and the maximum compressive strength obtained from dataset of 350 concrete cubes were 5 N/mm^2 and 55 N/mm^2 respectively. Initially non-destructive testing was carried out on the cubes followed by subjecting the specimen to compressive loading upto failure in compression testing machine (CTM). Schmidt RBH was used to determine the hardness of concrete cubes by measuring the RBH number. Pundit USV instrument was used to record the velocity of propagation (km/s) of pulses through the test specimen. The velocity of propagation indicates the quality of concrete. Finally Windsor penetrometer was employed to find the depth of penetration of the probe. The depth of penetration indicates the strength of in-situ concrete.

5. Results & Discussion

After the completion of experiment, we have four types of data available viz. depth of penetration (mm), pulse velocity (km/s), RBH number which has been obtained by performing NDT on the concrete cubes. The compressive strength of concrete is also available which has been found out by subjecting the cubes to compressive load upto failure in compression testing machine (CTM). The dataset associated with the SVR model is classified into input and output. The NDT results form the input data set and the compressive strength of concrete constitutes the output dataset. R^2 formed the basis of comparison of various models generated. The range of R^2 is 0-1 where 1 indicates 100% accuracy in prediction. Initially prediction models using single NDT results were generated. Later, 2 NDT input combination prediction models were produced. 3 input prediction models were also created. Along with SVR, predictions using ANN on single NDT techniques were also carried out [17] . From the results, it was concluded that the accuracy of prediction of SVR is higher than that of ANN. Table 2 shows the value of R^2 for ANN and SVR.

Table 2: Coefficient of determination (R^2) of ANN & SVR

NDT Technique	R^2 of ANN	R^2 of SVR
RBH	0.857	0.911
WPP	0.786	0.8299
USV	0.468	0.769

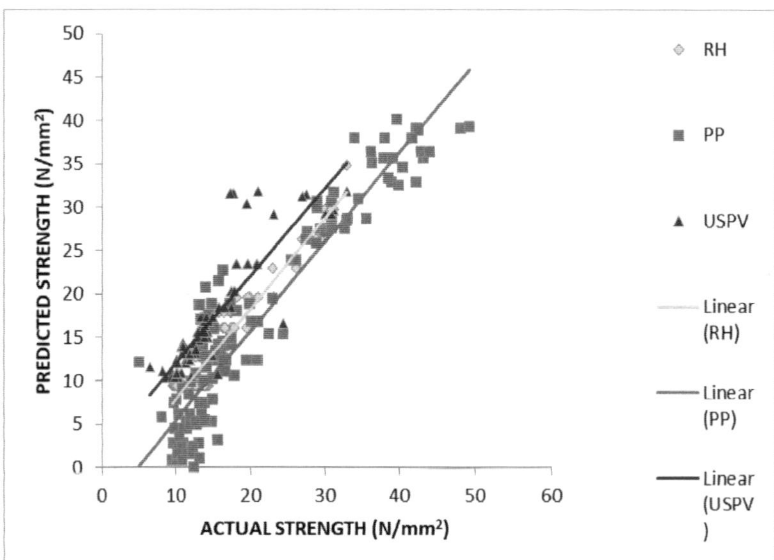

Figure 1: Graph depicting prediction results using 1 NDT technique.

Figure 1 shows the prediction result when single NDT result is used as input for the SVR model. From the graph we can observe the points are scattered from away prediction curve for USV and that of RBH is nearest. This implies that the accuracy of prediction of RBH is the best among the 3 NDT employed. The prediction result when 2 NDT results are used in combination for the SVR model is showed in figure 2. It can be observed that the extent of scatter is maximum for WPP-ultrasonic pulse velocity combination and minimum for RBH-WPP combination making it the most reliable prediction model combination when 2 NDT results are used in combination for concrete compressive strength prediction.

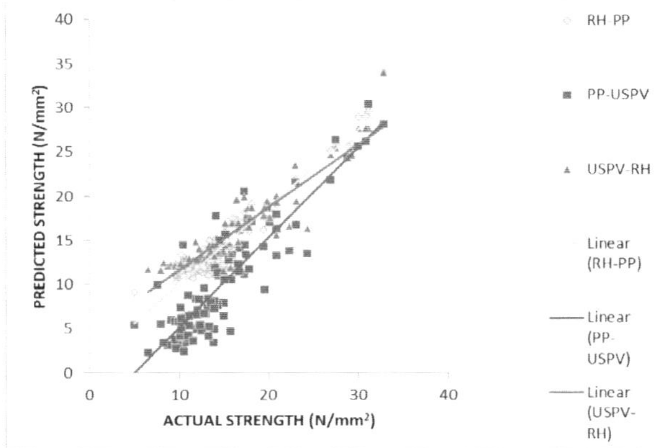

Figure 2: Graph depicting prediction results using 2 NDT technique combinations.

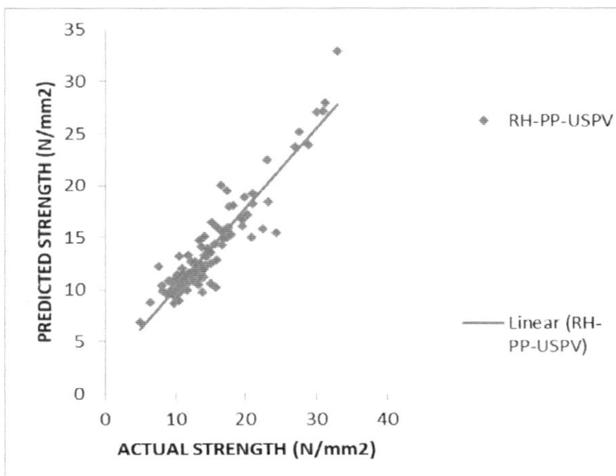

Figure 3: Graph depicting prediction results using 3 NDT technique combinations.

Figure 3 shows the plot between the actual and the predicted compressive strength of concrete when 3 NDT results are used as input for the SVR prediction model. Ideally when the input for the prediction model is more, higher accuracy must be obtained. But such a pattern is not observed here because of lower quality of results observed in USV technique. Table 3 illustrates the statistical parameters that were defined for various prediction models.

Table 3: Statistical parameters of various NDT techniques

NDT Technique	R^2	MAD	RMSE	MAPE
USV	0.769	5.117	7.291	14.862
RBH	0.911	3.518	4.284	10.084
WPP	0.8299	4.42	6.063	12.529
USV-WPP	0.832	4.314	5.121	14.586
WPP-RBH	0.91	3.605	4.382	9.891
USV-RBH	0.896	3.735	4.487	10.489
USV-WPP-RBH	0.899	3.36	4.309	8.488

From observing the results, it can be concluded that the use of USV is not feasible due to inferior accuracy of prediction. To have a superior accuracy of prediction using USV, we should have a larger input data set of USV.

6. Conclusion

In the study carried out, a prediction model was generated to predict the compressive strength of concrete using support vector regression which is a machine learning technique. The prediction modelling was executed in python language. It has been found out that RBH produces the best prediction output in comparison to other NDT techniques. The values of MAPE, MAD, RMSE and R^2 also support the findings. Furthermore the output of SVR was compared with ANN and SVR was found to be superior. Hence it can be used for accurate prediction.

Acknowledgement

The authors thank National Institute of Technology Kurukshetra for providing technical support for successfully conducting the research work.

References

[1]DeRousseau M A, Laftchiev E, Kasprzyk J R, Rajagopalan B and Srubar W V 2019 A comparison of machine learning methods for predicting the compressive strength of field placed concrete *Constr. Build. Mater.* **228,** 116661

[2]Ali-Benyahia K, Sbartaï Z M, Breysse D, Kenai S and Ghrici M 2017 Analysis of the single and combined non-destructive test approaches for on-site concrete strength assessment: General statements based on a real case-study *Case Stud. Constr. Mater.* **6,** 109–19

[3]Chou J S and Pham A D 2015 Smart Artificial Firefly Colony Algorithm-Based Support Vector Regression for Enhanced Forecasting in Civil Engineering *Comput. Civ. Infrastruct. Eng.* **30,** *715-32*

[4]Sobhani J, Khanzadi M and Movahedian A H 2013 Support vector machine for prediction of the compressive strength of no-slump concrete *Comput. Concr.* **11,** 337–50

[5]Pham A D, Hoang N D and Nguyen Q T 2016 Predicting Compressive Strength of High-Performance Concrete Using Metaheuristic-Optimized Least Squares Support Vector Regression *J. Comput. Civ. Eng.* **30,** 28–31

[6]Helal J, Sofi M and Mendis P 2015 Non-destructive testing of concrete: A review of methods *Electron. J. Struct. Eng.* **14,** 97–105

[7]Akkurt S, Tayfur G and Can S 2004 Fuzzy logic model for the prediction *of* cement compressive strength *Cem. Concr. Res.* **34,** 1429–33

[8]Reich Y 1997 Machine learning techniques for civil engineering problems *Comput. Civ. Inftrastruct. Eng.* **12,** 295–310

[9]Demir F and Armagan Korkmaz K 2008 Prediction of lower and upper bounds of elastic modulus of high strength concrete *Constr. Build. Mater.* **22,** 1385–93

[10] Salehi H and Burgueño R 2018 Emerging artificial intelligence methods in structural engineering *Eng. Struct.* **171,** 170–89.

[11] Yan K, Xu H, Shen G and Liu P 2013 Prediction of splitting tensile strength from cylinder compressive strength of concrete by support vector machine *Adv. Mater. Sci. Eng.* 2013.

[12] Taffese W and Sistonen E 2017 Machine learning for durability and service-life assessment of reinforced concrete structures: Recent advances and future directions *Autom. Constr.* **77,** 1–14.

[13] Çevik A, KURTOĞLU A E, Bilgehan M, Gülşan M E and Albegmprli H M 2015 Support vector machines in structural engineering: A review *J. Civ. Eng. Manag.* **21,** 261–81

[14] Chang C and Lin C 2013 LIBSVM : A Library for Support Vector Machines 1–39

[15] Pedregosa F, Grisel O, Weiss R, Passos A and Brucher M 2011 Scikit-learn: Machine Learning in Python **12,** 2825–30

[16] Oliphant T E 2007 Python for scientific computing *Comput. Sci. Eng.* **9,** 10–20

[17] Singh V P and Kotiyal Y C 2013 Prediction of compressive strength of concrete using artificial neural networks *Res. J. Pharm. Technol.* **10,** 35–40

Surface Characteristics Enhancement of Biocompatible Mg Alloy AZ31B by Cryogenic Milling

Subham Chhetri[1*], Mohammad Tariq[1], Stephen Dilip Mohapatra[1], V Vikali Sumi[1], Alomi P Zhimomi[1], Rahul Davis[2*], Abhishek Singh[2]

[1]Department of Mechanical Engineering, Vaugh Institute of Agricultural Engineering and Technology, SHUATS, Prayagraj - 211007, India
[2]Department of Mechanical Engineering, National Institute of Technology Patna,
Patna - 800005, India
Email: [1*]shoebumgohan@gmail.com [2*] rahul.me18@nitp.ac.in

Abstract: The mechanical, physical and chemical properties of Magnesium alloys have undoubtedly proven compatible and efficient for their aid in biomedical sector. Magnesium tends to corrode rapidly in in-vivo environment limiting its efficiency and life expectancy as a biomedical implant. This has led to the study and research of the properties of these excellent materials in order to increase its corrosion resistivity property. In the recent years through thorough research and study it has been proven that cryogenic machining is an efficient technique for surface integrity improvements which plays a vital role in the increment of corrosion resistivity of a material. This work in particular is one of the studies in which the milling operation is carried out on a CNC milling machine on the commercially available AZ31B Magnesium alloy by a cemented carbide tool under two different milling environments namely wet and cryogenic. Different machining factors were opted and comparative investigation was conducted between the two machining conditions. The post machined results were further compared for minimum surface roughness. The comparative analysis presented a better surface finish in case of cryogenic milling than wet milling.

Keywords: Magnesium alloy AZ31B; cryogenic treatment; wet and cryogenic machining; surface integrity; corrosiveness.

1. Introduction

Biomaterial is any material which can replace any damaged parts of a body for a certain period of time or permanently. There has been a great development in the field of biomedical industry regarding biomaterial. Materials like polymeric biomaterial has been used in applications such as bone tissue engineering due to their excellent formability [1]. Metal implant have great mechanical properties, thus used for repairing of bone fracture. Stainless steel, Co and Ti based alloys are one of most well-known implants [2]. Ceramics, too are been used for coating purposes because to their non-toxicity and biocompatibility like properties [3]. Each biomaterial serves specific applications [4, 5]. The specific features of some of them make their machining quite challenging. One machining process that would work on one material may not work on another material or maybe one [1]material would possess contrasting requirement than the other [6, 7]. Because of the lack in the advancement of technology and resources, the surgical operation was merely limited to focus intervention without the knowledge of possibility of aid of biomedical implants. But with the passing of time and technological breakthrough, better perception of biology and human anatomy was made possible. This ultimately paved its way to the design and development of bio medical implants that were biocompatible and served great purpose [8]. One of the commonly used degradable biomaterials is magnesium and its alloys. These materials possess great potential and excellent biocompatible properties for manufacturing biomedical implant. Therefore, there are various machining processes involved in the fabrication of magnesium (Mg) biomedical implants which vary according to the specific applications and requirements [9]. Despite its bright scope, pure magnesium is reactive in nature and subsequently corrosive. This limitation can be overcome by alloying pure Mg. Alloying is a method where various metals in varying concentrations are added to improve and boost the mechanical properties such as ductility, strength and corrosion resistivity of pure Mg. Alloys of Mg can be ternary, binary or more. The most popular metal used as Mg alloy are aluminium (Al), Rare Earth Element (REE) [10], Zirconium (Zr) and Zinc (Zn). Some other alloying components are Ca, Mn, Fe, Si, Bi, Sr, Li, Ag etc. The use of Mg alloy as an implant dates back to 1878, where it was used in cardiovascular application as wire and another ligature. It is also used as a stent. A stent is an artificial tubular device that is surgically inserted into a natural passage in the body to counteract or prevent it from clogging [11]. In orthopedic specifically in traumatology as pegs, fixation plates, and screws to secure bone fractures. Magnesium alloy has the advantage of being compatible to mechanical and physical properties to that of human bone. Magnesium is naturally found in human bone composition and is a required metal for metabolism [12]. However due to its corrosive nature its wide potential is dimmed. Extensive studies and research are taken up to overcome these challenges. AZ31B is one of the commercially available magnesium alloys, which is composed of two alloying materials namely aluminium and zinc. These materials have their own advantages and their alloying adds to the property of Magnesium. Aluminium elevates strength and hardness. It makes the alloy easier to cast as it widens the melting range. With aluminium content greater than 6%, the alloy is heat treatable. On the other hand, studies show that 6% zinc in Mg improves the corrosion resistance properties. Also, Zn is known as one of the frequent elements in alloying of Mg [13]. It is also one of the most common mg alloys after aluminium. Bio functional trace metallic elements such as Ag, Ca, Sr and Cu when alloyed with Mg, promotes bone cell activation and also quickens the bone formation. In case of Machining of bio materials like AZ31B, optimization of factors in milling operation like feed rate, cutting speed and depth of cut play a very vital role in the increment of process productivity and overall machined biomaterial quality [14]. A study suggested that the cryogenic machining successfully imparted grain improvement on Mg alloys while ensuring improvement of the same material's corrosion resistance. The study was performed to explore AZ31B's micro structural modifications in dry and cryogenic machining [15]. Another study attempted to investigate the various influence of machining factors like cutting rpm and feed rate on various output variables such as surface roughness, microstructure, generated forces and temperature. A post machined comparison was made on the two different milling environments. The comparative investigation proved that the combination of low feed rate and high cutting rpm under cryogenic milling environment portrayed better surface finish than that of dry [16]. An investigation was performed on the behaviour of Mg Alloy AZ31B in cryogenic machining. Different mechanisms of alteration in the microstructure and severe plastic deformation (SPD) were examined on the basis of various reviews of experimental results found in literatures. It was found that the combination of cryogenic machining with large edge radius tool proved to improve surface integrity of the material [17]. The effect of dry and cryogenic machining with the combination of different tools carrying different cutting-edge radius on the surface integrity of Mg Alloy AZ31B, was executed in one research. It was observed that the combination of cryogenic machining with large edge tool radius displayed improved surface integrity properties with regard to better surface finish, remarkable grain refinement, and larger intensity of basal plane on the machined surface and ten times larger compressive stress profiles significantly enhancing the functional performance of the sample [18].

2. Methodology

2.1 Material

The Material used for the experiment was Mg Alloy AZ31B with dimension 50 x 50 x 20 mm. The chemical composition and the mechanical properties of the material are illustrated in Table 1.

Table 1: Chemical composition and mechanical properties of Mg Alloy AZ31B

Al (wt%)	Zn (wt%)	Mn (wt%)	Cu (wt%)	Si (wt%)	Fe (wt%)	Mg (wt%)	Tensile Strength (MPa)	Density (g/cm^3)	Thermal Conductivity (W/m K)
2.7	0.9	0.4	0.006	0.018	0.002	Balance	260	1.78	96

2.2 Experimental Setup

A MTAB XL CNC milling machine of bed size 360×32mm was used for the experimental work. A coated carbide end mill tool (KYOCERA SER 1M Ti-NAMITE) of 6 mm diameter was used as a cutting tool for each machining trial. The experimental setup and pictorial view of the machining process are shown below in Figure 1. A single experiment consisted of producing of a circular slot on the workpiece

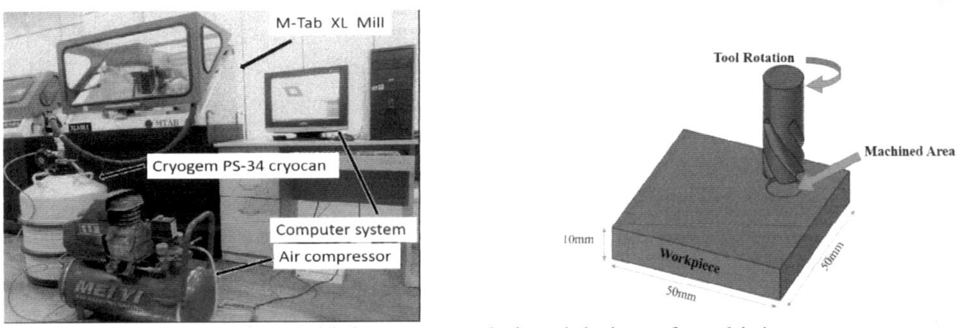

Figure 1: Cryogenic machining setup and pictorial view of machining process

The design includes the proper selection of variables or factors and their relation with the surface. For this particular experiment one-factor-at-a-time (OFAT) method was used to see the effect of variation in each factor at a time on the surface characteristics namely surface integrity and crack formation on the machined region. Number of cracks can act as the indicator of corrosiveness of the machined surface, since it was found from the studies that better the finish of the machined surface better will be the corrosion resistance [19].
In this method, an experiment takes place while keeping one factor as constant and others as variables at a time. The chosen factors of interest for the present work were milling environments, cutting speed (v_c), feed rate (f) and depth of cut (a_p). Wet machining was done via Zed plus cutting oil while cryogenic machining was done via liquid nitrogen (LN_2) at a temperature of -196°C. Table 2 shows the process factors for the proposed milling operation of AZ31B.

Table 2: Factors and their corresponding levels

Factor	Level 1	Level 2
Milling environment	Wet	Cryogenic
Cutting Speed (rpm)	1400	2800
Feed Rate (mm/rev)	80	120
Depth of Cut (mm)	0.4	0.8

Figure 2: Close view of Spray nozzle for (a) wet and (b) Cryogenic milling, and (c) workpiece sample with some circular shaped machined surfaces

3. Results and Discussion

Table 3 shows the results obtained for the surface roughness (SR) under different process factor value combinations. Each experimental trial was run twice for each set of process factors and is represented by T1 and T2, respectively. The roughness measured in each trial of the machined surfaces were measured with the help of a surface roughness tester (Model: Precise-TR110P) followed by averaging of both the obtained values. It was observed that each factor had its significant effect on the surface roughness of AZ31B. A bar graph along with the standard deviation to indicate the measure of scatter of results (since an average value was taken) was plotted [20]. It can be observed through Figure 3 (a) that when feed rate and depth of cut were kept constant at 80mm/rev and 0.4mm respectively, surface roughness decreases with increase in cutting speed. This can be the result of less cutting forces and thrust forces produced at higher cutting speed which ultimately reduces tool vibrations resulting in incremented surface finish. In addition to this at higher cutting speeds, the formation of built-up edges is also less promoting superior quality of surface [21, 22]. A study revealed that there are some locations which remains uncut when feed rate is kept higher and amount of small portions of unmachined material area increases. This is because of the helical tool path of the cutting tool [23, 24]. Thus, at a feed rate of 80 mm/rev the SR was much better (Figure 3 (b)) then is case of 120 mm/rev.

Table 3: Experimental results

S.N.	Spindle Speed (v_c) (rpm)	Feed rate (f) (mm/rev)	Depth of cut (a_p) (mm)	Surface Roughness (µm) Milling environment					
				Wet			Cryogenic		
				T1	T2	Avg.	T1	T2	Avg.
1	1400	80	0.4	0.58	0.60	0.59	0.54	0.56	0.55
2	1400	120	0.4	0.79	0.81	0.80	0.75	0.71	0.73
3	1400	80	0.8	0.99	1.03	1.01	0.96	0.94	0.95
4	1400	120	0.8	1.21	1.23	1.22	1.11	1.15	1.13
5	2800	80	0.4	0.53	0.49	0.51	0.48	0.50	0.49
6	2800	120	0.4	0.71	0.73	0.72	0.69	0.65	0.67
7	2800	80	0.8	0.92	0.94	0.93	0.90	0.88	0.89
8	2800	120	0.8	1.15	1.13	1.14	1.06	1.08	1.07

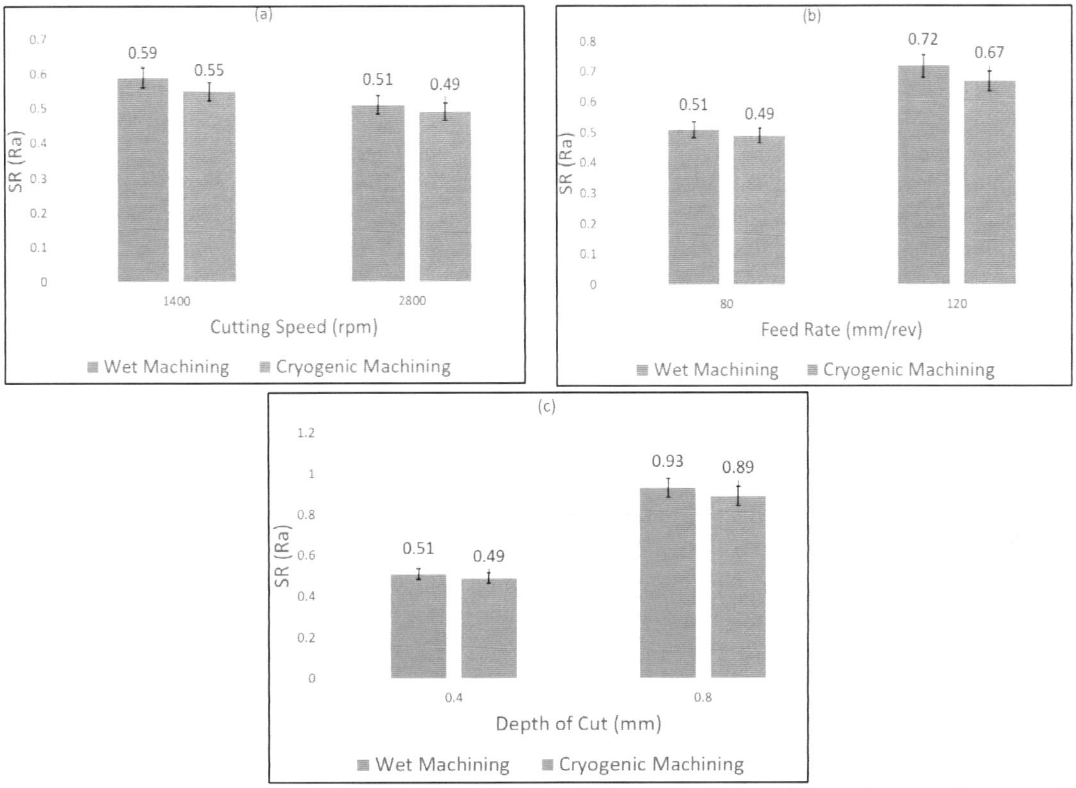

Figure 3: Comparison between surface roughness and (a) cutting speed (b) Feed rate and (c) Depth of cut

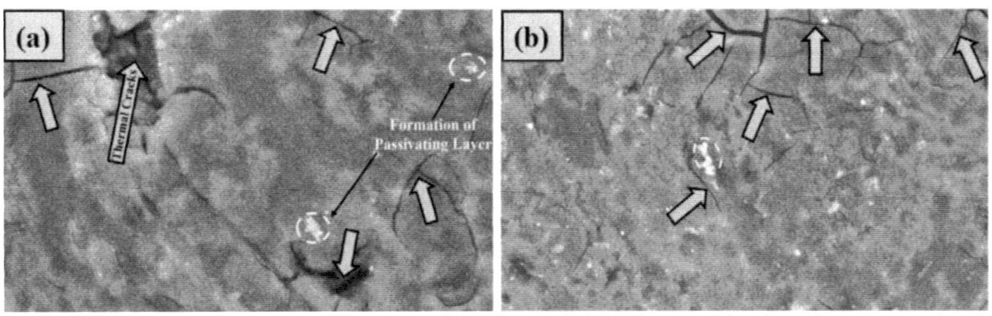

Figure 4: SEM images (at a magnification of 1500 ×) of (a) Wet machined surface by end mill (b) Cryogenically machined surface by end mill

Figure 3 (c) shows a comparison between wet and cryogenic machining with variation in depth of cut of 0.4 mm and 0.8mm and constant cutting speed and feed rate of 2800 rpm and 80 mm/rev respectively. It is observed that at lower depth of cut the surface roughness is low. This could be the result of higher cutting force and higher generation of friction as the tool cuts deeper giving higher surface roughness. The best result was found in case of high cutting speed and low feed rate under cryogenic condition. It can be observed from Figure 3 (a), (b) and (c) that cryogenic machining achieved lower values of the surface roughness in each case. This could be for the reason that under cryogenic machining condition the heat generated during machining is very low as compared to a normal wet machining. It also increases the surface hardness. It also helps in obtaining a better corrosion resistance [25]. Cutting force and thrust force were also less in case of cryogenic machining which reduces the tool wear, hence decreases SR.

The above portrayed Figure 4 represents the SEM images of the machined surface under two different machining conditions. As illustrated above, image (a) displays machined surface by end mill under wet condition. Likewise, image (b) displays machined surface by end mill under cryogenic condition. A number of cracks were observed on the surface of the workpiece. These cracks are known as thermal cracks. Due to higher temperature and increased cutting force the number of cracks is more in case of wet machined surface by end mill [26]. These cracks were observed where a passivating layer was less. Passivation incorporates formation of an external layer that is enforced as a micro coating which because of the chemical reaction with the base material results in its formation. Passivation formation may also be a result of impromptu oxidation that takes place in atmosphere.
It was observed that the passivating layer was much more well spread in case of cryogenically machined surface. Not all but some alloys of magnesium can establish a protective surface layer or film in exclusive solution or condition [27]. This layer protects the surface from getting corrode before time. However, it will not completely stop the corrosion because it also has small gaps between them [28].

4. Conclusion

The present paper has displayed a relation among different factors and surface roughness under wet and cryogenic machining of AZ31B magnesium alloy. Following conclusion has been made through these experiments:

- Cryogenic milling has been proven to be a better cooling alternative (than wet milling environment) for obtaining better surface finish with higher precision.
- In cryogenic milling the surface roughness was reduced to around 20% over wet machining.
- The passivating layer act as protecting layer for the surface from the corrosion.
- Experiments revealed that rise in cutting speed (v_c) values, and decrease in feed rate (f) and depth of cut (a_p) can produce minimum surface roughness.

References

[1] Ratner B D 2005 *Biomaterials Science: An introduction to materials in medicine*, **26**, 5093
[2] Lodhi M J K, Deen K M, Greenlee-Wacker M C, Haider W 2019 Additively manufactured 316l stainless steel with improved corrosion resistance and biological response for biomedical applications, *Addit. Manuf.*
[3] Tahmasebifar A, Kayhan S M, Evis Z, Tezcaner A, Cinici H., Koc M 2016 Mechanical electrochemical and biocompatibility evolution of az91d magnesium alloy as a biomaterial, *J. Alloys Compd.*, **687,** 906–919
[4] Singh D, Singh R, Boparai K S, Farina I, Feo L, Verma A K 2018 In-vitro studies of ss 316 l biomedical implants prepared by fdm, vapor smoothing and investment casting, *Compos. B. Eng.*, **132**,107–114
[5] Raghavendra G M, Varaprasad K, Jayaramudu T 2015 Biomaterials: nanotechnology Applications for Tissue Engineering **21**–44, 2015.
[6] Ratner B D 2013 Introduction: biology and medicine – key concepts in the Use of biomaterials in surgery and medical devices *Biomater., Sci.* 393–394
[7] Park J, Lakes R S 2007 Biomaterials: an introduction 328-335
[8] Chen J, Tan L, Yu X, Etim I P, Ibrahim M Yang K 2018 Mechanical properties of magnesium alloys for medical application: a review, *J. Mech. Behav. Biomed.*, **87,** 68–79
[9] Maji P K, Banerjee A J, Banerjee P S, Karmakar S 2014 Additive manufacturing in prosthesis development—a case study, *Rapid Prototyp. J.*, **20**, 480–489
[10] Li Z, Gu X, Lou S, Zheng Y 2008 The development of binary mg–ca alloys for use as biodegradable materials within bone *Biomaterials,* **29**(10), 1329–1344
[11] Sillekens W H and Bormann D 2012 Biomedical applications of magnesium alloys **13,** 427-452
[12] Saris N E L, Mervaala E, Karppanen H, Khawaja J A, Lewenstam A 2000 Magnesium: an update on physiological, clinical and analytical aspects *Clin. Chim. Acta*, **294** (1-2), 1–26
[13] Zhang S, Zhang X, Zhao C, Li J, Song Y, Xie C, Jiang Y 2010 Research on an Mg–Zn alloy as a degradable biomaterial *Acta Biomater.* **6**(2), 626–640
[14] Shi K, Zhang D, Ren J 2015 Optimization of process factors for surface roughness and microhardness in dry milling of magnesium alloy using taguchi with grey relational analysis *Int. J. Adv. Manuf. Tech.*, **81**(1-4), 645–651
[15] Hussein R O and Northwood D O 2014 Improving the performance of magnesium alloys for automotive applications, high performance and optimum design of structures and materials **137,** 531-543

[16] Balan A S S, Tibrewal V, Dak K, Himanshu A, Kumar H, Kuppan P 2019 Cryogenic machining of az31b magnesium alloy for bio-implant applications, advances in forming, machining and automation 239-251
[17] Shen N, Ding H, Pu Z, Jawahir I S, Jia T 2017 enhanced surface integrity from cryogenic machining of mg alloy az31b: a physics-based analysis with microstructure prediction *J. Manuf. Sci E-T Asme,* **139**(6), 061012
[18] Pu Z, Outeiro J C, Batista A C, Dillon O W, Puleo D A, Jawahir I S 2012 Enhanced surface integrity of mg alloy az31b by cryogenic machining towards improved functional performance of machined components *Int. J. Mach. Tool. Manu.,* **56,** 17–27
[19] Walter R, Kannan M B 2011 Influence of surface roughness on the corrosion behaviour of magnesium alloy, *Mater. Des.,* **32,** 2350–2354
[20] Zagórski I and Korpysa J 2019 Surface quality in milling of az91d magnesium alloy *ASTRJ,* **13**(2), 119-129
[21] Danish M, Ginta T L, Habib K, Carou D, Rani A M A, Saha B B 2017 Thermal analysis during turning of az31 magnesium alloy under dry and cryogenic conditions *Int. J. Adv. Manuf. Tech.,* **91,** 2855–2868
[22] Pu Z, Song G L, Yang S, Outeiro J C, Dillon Jr O W, Puleo D A, Jawahir I S, 2012 Grain refined and basal textured surface produced by burnishing for improved corrosion performance of az31b mg alloy *Corros. Sci.,* **57,** 192–201
[23] Malleswara Rao J N, Sumalatha M, Kesava Rao V V S, Anurupa V, Srivalli G 2016 Variation of surface roughness with feed rate on mild steel components produced by cnc lathe *Int. Res. J. Eng. Technol.,* **3**(06)
[24] Mebrahitom A, Rizuan D, Azmir M, Nassif M 2016 Effect of high-speed milling tool path strategies on the surface roughness of stavax esr mold insert machining *IOP Conf. Ser.Mater. Sci. Eng.,* 114
[25] Jawahir I S, Puleo D A, Schoop J 2016 Cryogenic machining of biomedical implants materials for improved functional performance, life and sustainability *Procedia CIRP,* **46,** 7 – 14
[26] Jia W, Ma L, Le Q, Zhi C, Liu P 2019 Deformation and fracture behaviour of az31b mg alloy at elevated, temperature under uniaxial compression *J. Alloys Compd.,* **783,** 863-876
[27] Cao F, Song G L, Atrens A 2016 Corrosion and passivation of magnesium alloys *Corros. Sci.,* **111,** 835-845
[28] Ijiri M, Shimonishi D, Tani S, Okada N, Yamamoto M, Nakagawa D, Tanaka K, Yoshimura T 2019 Improvement of corrosion resistance of magnesium alloy by high- temperature high-pressure cavitation treatment *International Journal of Lightweight Material and Manufacture,* **2,** 255-26

Reduction of positional error in a path of a 2 DOF serial planar manipulator

R N Tatte[1*], H T Thorat[2] and P V Deosant[3]

[1,3]PhD Research Scholar, Visvesvaraya National Institute of Technology, Nagpur, MH

[2] Professor, Visvesvaraya National Institute of Technology, Nagpur, MH

[*]E-Mail: rupeshtatte0096@gmail.com

Abstract: Tracing a path is one of the major applications of a robotic arm. In this paper, a 2 DOF serial planar manipulator which traces the desired path is discussed. Quite often the path is defined by accuracy points lying on it. To trace these points, joints of links of the manipulator are required to be given as a coordinated inputs θ_1 and θ_2 (input angles of two links). This path is traced with a desired speed indicated by the time at which the accuracy points are reached.

In open-loop control, this is achieved by giving time, voltage/current input to driving motors. This path is determined based on the torque, acceleration, velocity of the linkages and individual joints. Hence the relationship between time and joint angles is necessary. When joint angles θ_1 and θ_2 are interpolated between two accuracy points, precise coordinated motion is not possible. Some deviation from the desired path will occur. This is the positional error for tracing the path. Here, analysis has been done to find the effect of degree of interpolation on the positional error.

Keywords: joint angles; path; positional error; spline interpolation.

1. Introduction

In industry, to perform high precision work robotic arm is used. One of the major applications of a robotic manipulator is to accurately follow a desired path from initial to last position. To get the desired position, joint motors are to be controlled by controlling the current. How much current is required with time is dependent on torque and this torque is known from the relationship of joint angles[1]. To accomplish this, robotic manipulators are commanded with program sequences that are executed in digital computers. Within these computers, there is an operating software that provides the information on positions and orientations of the end effectors by computing them in the joint coordinated system as function of time. Generally, whole path is generated by using single polynomial and it is used in most of the computers commanded industrial robots which are not exact enough such that difference between the actual and the desired position can be significant. Differences as high as 10 mm may be possible for the robots with roughly 0.1 mm repeatability[2]. If the function used of lower degree polynomial, the difference goes high. To decline this positional error, higher degree polynomial as a function of joint variable is recommended. However, the complex path cannot be executed by the single polynomial even if higher degree polynomial is used[3]. At the same time computer needs to handle more complex calculations and requires more storage data.

To overcome this complexity, whole path is being composed piece by piece known as a spline. By interpolating each spline, a continuous and smooth curve is made out of it of the required degree[4]. For interpolating each spline, joints of the links are required to be given as a coordinated input θ_1 and θ_2. Positions of joint angles have been identified using inverse kinematics of 2 DOF serial planar manipulator[5]. After accuracy points to obtained continuous motion, the relationship of joint angles with respect to time is required to be given. To start with the position of the accuracy points and the time interval between them is known, the relationship between time and joint angles are determined independently by interpolation. Such a method does not guarantee coordinated motion and hence the possibility of positional error that is a deviation from the desired path exists. An attempt has been made to use different degrees of interpolation to determine the effect of positional error.

2. Path generation using spline

Whenever end-effector moves from one point to another point, the motion is defined either as from point to point or as path generation. This path is generated with a specific objective which can be like avoiding the obstacles, optimizing to minimize energy/power requirement/execution time/jerk and so on. These are the criteria by which paths are identified [6].

Path can be defined by an equation or by accuracy points lying on the desired path. These equations are possible for a simpler path but they become complex for the intricate path. If there is a standard path, it is sufficient to give only one equation, however for a complex path it may not be sufficient to define it with only one equation. For this, it is required to use combinations of equation of required degree [7],[8],[9].

Generating coordinated inputs based on equations becomes complex under such situations. Using accuracy points is a simpler method. Using these accuracy points, coordinated inputs are obtained using inverse kinematics of 2 DOF serial planar manipulator. These positions are used for interpolation using 'spline' and continuous path is obtained[4]. Depending on the degree of spline used, path accuracy varies, however it is accurate at the defined point. Inaccuracies because of the use of spline to defined path are termed here as path inaccuracies and in terms of error, it is called as positional error (e_{pd}).

3. Generating path in time domain

Controlling the speed of an end-effector, while tracing a path, is also an important issue. Generating a specific path involves only the relationship between θ_1 and θ_2 and precise calculations of θ_1 and θ_2 is feasible for all the points defined on the path. However, if the path is defined with time, θ_1 and θ_2 are also required to be defined with reference to time and hence now the relationship between θ_1 and θ_2 is also required to be defined with reference to time. If the time verses θ_1 and time verses θ_2 curves are to be defined with reference to the accuracy points by some kind of interpolation, this is likely to cause deviation from the planned path. This deviation from the planned path is called as positional error (e_{nc})[5]. Here it is called positional error due to non-coordination. Defining path by accuracy points and hence identifying time versus joint angle relationship based on the accuracy points can work as a generalized method for controlling joint motor motion.

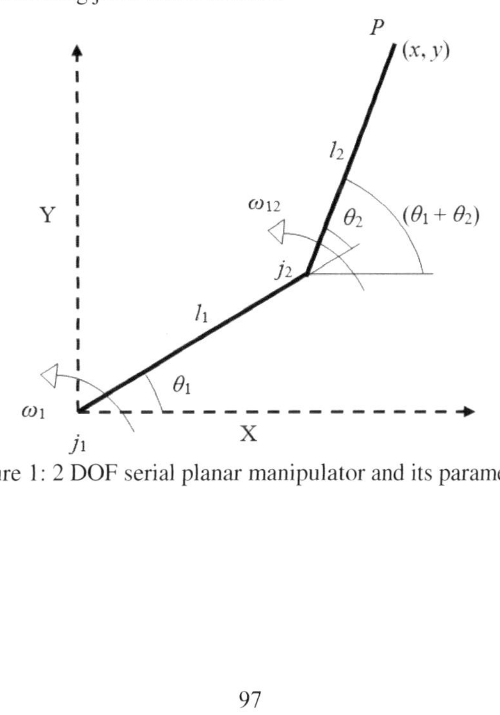

Figure 1: 2 DOF serial planar manipulator and its parameter.

Fig.1 shows the kinematic arrangement of 2 DOF serial planar manipulator where θ_l is the angle between link 1 with respect to horizontal axis, θ_2 is the angle between link 2 with respect to link 1. Similarly, (l_1, l_2) and (ω_1, ω_{12}) are the respective length and angular velocity of link 1 and link 2. 'P' is the position of end-effector with x and y coordinate

Equations used for direct and inverse kinematics are as under [10].

$$x = l_1 \cos(\theta_1) + l_2 \cos(\theta_1 + \theta_2) \tag{1}$$

$$y = l_1 \sin(\theta_1) + l_2 \sin(\theta_1 + \theta_2) \tag{2}$$

$$\theta_1 = \tan^{-1}(y/x) \pm \sin^{-1}\left(\frac{l_2}{\sqrt{x^2+y^2}} \sin\left(\cos^{-1}\left(\frac{l_1^2 + l_2^2 - x^2 - y^2}{2l_1 l_2}\right)\right)\right) \tag{3}$$

and

$$\theta_2 = \cos^{-1}\left(\frac{x^2 + y^2 - l_1^2 - l_2^2}{2l_1 l_2}\right) \tag{4}$$

To get the equations of joint angle θ_1 and θ_2 with respect to time through interpolation, it is required to specify the positions of θ_1 and θ_2 with respect to time which has been obtained from the known accuracy points (x, y) lying on the desired path. From these equations, it is possible to get θ_1 and θ_2 at different positions at a different time. Since these joint angles are a function of time (t), it is easy to find out the successive derivatives of the equation. The relationship of a joint coordinated system with respect to time has been obtained from the given polynomial equation.

$$\theta(t) = a_n t^n + a_{n-1} t^{n-1} + \cdots + a_1 t + a_0 \tag{5}$$

$$\omega(t) = n a_n t^{n-1} + (n-1) a_{n-1} t^{n-2} + \cdots + a_1 \tag{6}$$

Eqs. (5) and (6) is the 'n' degree polynomial equation for joint angle and joint angular velocity. To use these equations, it is required to calculate all (n+1) constants in the polynomial. These constants have been calculated from the initial and final boundary conditions of each segment. The number of initial and final boundary conditions required are depended on the number of segments that are generated by splitting the path and also on the degree of the polynomial.

4. Estimation of positional error

In this, two different paths have been analysed which is a straight-line path and parabolic path for estimating positional error. To eliminate the effect of e_{pd}, single equation curves are used. Hence here forth, positional error e and positional error due to non-coordination e_{nc} are same (e = e_{nc}). To defined the time verses θ relationship, interpolation between the accuracy points has been done using three different degrees of interpolation linear, quadratic and cubic. Based on these three interpolations the angles are identified for ten segments. Using direct kinematics, the position of end-effector is determined and this is called as the actual path. The end conditions used for the three cases are as under.

Table 1: End conditions used for interpolation in a straight-line path.

θ_1 (degree)	θ_2 (degree)	ω_1 (deg./sec)	ω_2 (deg./sec)
-36.86	90	0	0

-34.26	100.80	2.60	10.80
-29.27	109.47	4.98	8.66
-21.90	115.94	7.36	6.47
-12.41	120.00	9.49	4.05
-1.41	121.38	10.99	1.38
10.20	120	11.62	-1.38
21.69	115.94	11.48	-4.05
32.64	109.47	10.95	-6.47
43.05	100.80	10.40	-8.66
53.13	90	0	0

Table 2: End conditions used for interpolation in parabolic path.

θ_1 (degree)	θ_2 (degree)	ω_1 (deg./sec)	ω_2 (deg./sec)
-36.86	90	0	0
-34.59	100.86	2.27	10.86
-30.77	109.95	3.82	9.08
-25.51	117.61	5.25	7.66
-18.89	124.05	6.61	6.44
-11.01	129.40	7.88	5.34
-1.97	133.72	9.03	4.32
8.03	137.07	10.01	3.34
18.79	139.46	10.76	2.38
30.04	140.89	11.24	1.43
41.49	141.37	0	0

For linear interpolation, initial and final joint angles are used in end conditions. For quadratic interpolation, initial and final joint angles, as well as initial angular velocities, are used in end conditions. Whereas in cubic interpolation, all the four boundary conditions are used. Straight-line path used in this analysis starts with the coordinates x=0.5 m, y=0 m and ends with x=0 m, y=0.5 m, whereas parabolic path starts with x=0.5 m, y=0 m and ends with x=0 m, y=0.25 m. Length of the links are taken as l_1 = 0.4 m and l_2 = 0.3 m.

The relationship between θ_1 and θ_2 for a segment is determined by linear interpolation. This has given 10 values of θ_1 and θ_2 and from these value 10 actual positions of end-effector are determined. The maximum positional error between the actual path and desired path for each segment is determined by taking perpendicular distance between the paths. A graph is plotted showing actual path and desired path. Further, the positional error in each segment is expressed as percentage positional error by taking segment length as reference length. This figure is also plotted segment-wise. Same procedure is repeated for all the three interpolation that is linear, quadratic and cubic interpolation. Same procedure is adopted for straight-line and parabolic path.

Figure 2 gives the actual path and desired path superimposed over each other for straight-line path and figure 3 gives it for parabolic path. Figure 4 and 5 give enlarge view for one segment each. All these figures represent that the positional error is very small. Figures 6 and 7 show percentage positional error for each segment for straight-line and parabolic path respectively. For straight-line path, linear interpolation gives positional error around 2.5 % uniformly all over the segment. However, in quadratic interpolation, there is a slight increase in positional error. However, cubic interpolation gives substantially low positional error around 1 %. In case of parabolic path, the error continuously decreases from 2 % to 1.4 % from first segment to the last segment for the linear interpolation. Quadratic interpolation has a similar trend but positional error is marginally low. However, the cubic interpolation has given the positional error around 0.6 %.

IOP Conf. Series: Materials Science and Engineering 1004 (2020) 012012 doi:10.1088/1757-899X/1004/1/012012

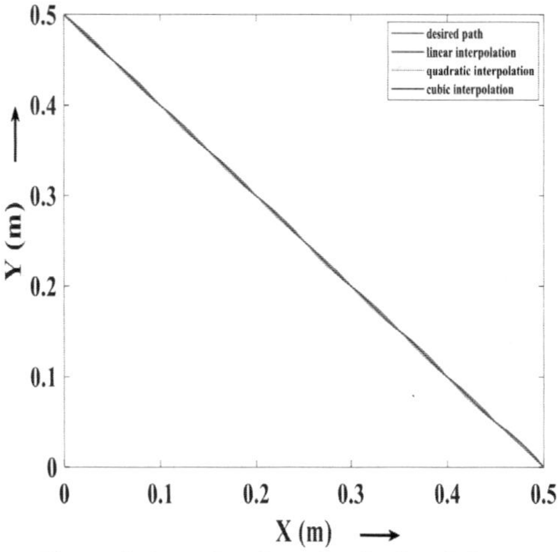

Figure 2: Actual path vs. Desired path for for straight-line path

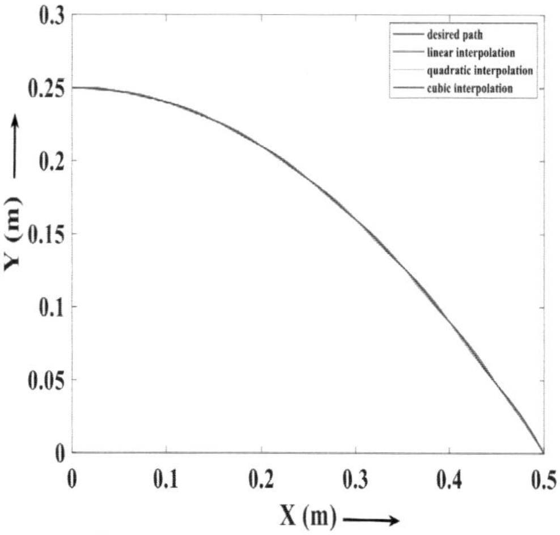

Figure 3: Actual path vs. Desired path parabolic.

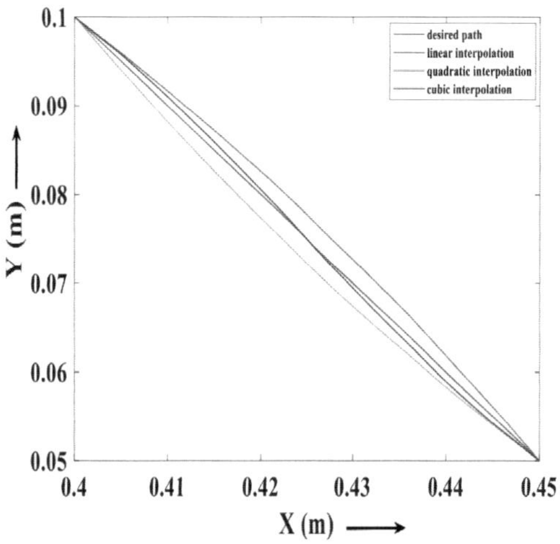

Figure 4: Enlarge view of one segment for straight-line path.

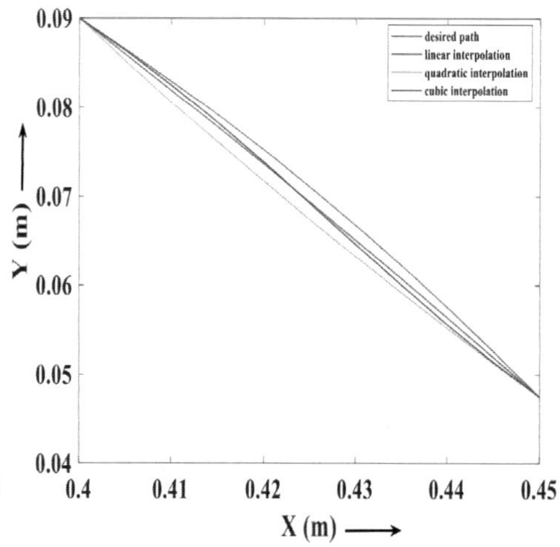

Figure 5: Enlarge view of one segment for parabolic path.

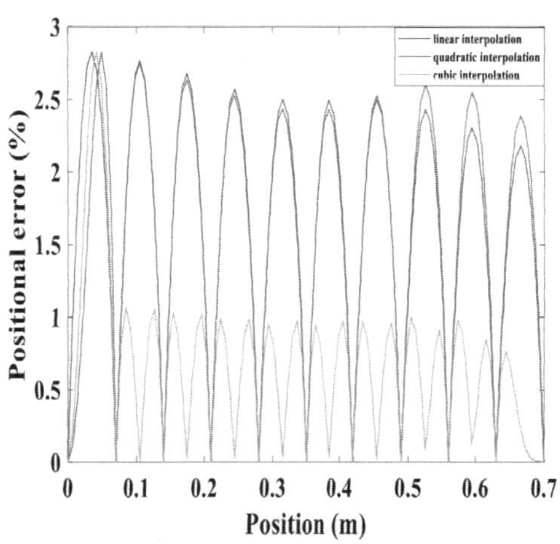

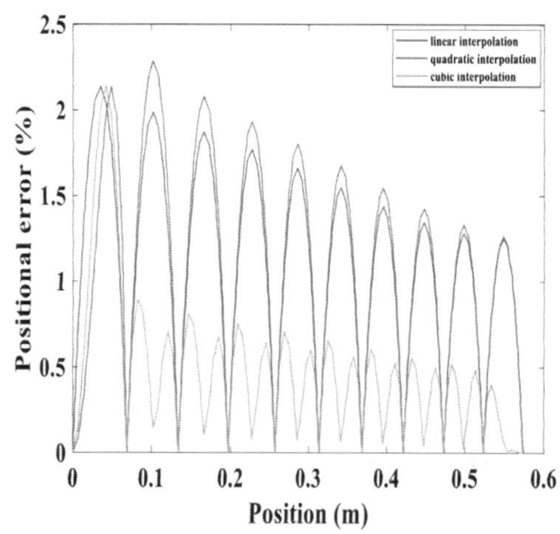

Figure 6: Positional error vs Position on path for straight-line path.

Figure 7: Positional error vs Position on path for parabolic path.

5. Method to reduce positional error in first segment

It is observed that, cubic interpolation gives nearly three times higher positional error in first segment compared to succeeding segments, whereas positional error in linear and quadratic interpolation is at the same level as that of other segments. This is caused due to undefined slope in the initial boundary condition. This definition does not ensure that path starts in the right direction. While defining the path if slope of the curve at the beginning is not defined or considered as 0 degree/sec. such errors are expected for the initial condition.

To take care of these errors a simple method is adopted. One additional segment is created at the beginning of the path having very short length (one tenth of the first segment). The slope of the first two points of the segment from zero has been considered as an initial boundary condition and θ_1 and θ_2 are determined by the equations used for interpolation.

Figure 8 and 9 shows the effect of this additional procedure in the first segment. Though there is no difference in linear interpolation, quadratic interpolation has given very large reduction in positional error about 90 % in the first segment only and other segments there is no effect. In case of cubic interpolation, there is a drop in positional error about 30 % in the first segment but still, it is higher than the remaining segment.

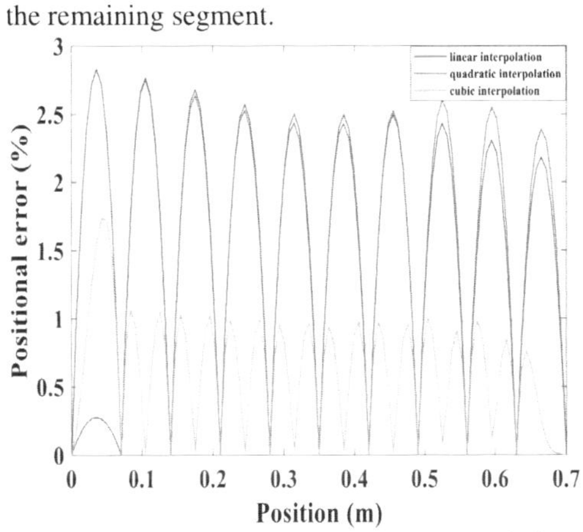

Figure 8: Positional error vs Position on path for straight-line path.

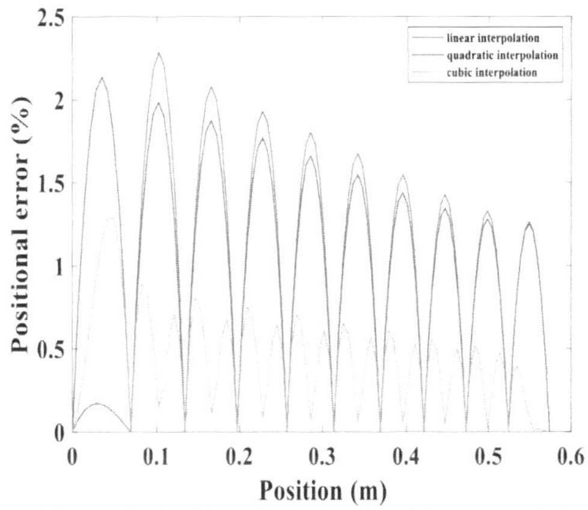

Figure 9: Positional error vs Position on path for parabolic path.

6. Conclusion

It is observed that the interpolation of θ_1 and θ_2 between accuracy points gives positional error in a path, this is true for a straight-line path (fist-degree curve) and also for parabolic path (second-degree curve). In both the profile, linear interpolation has given a large positional error in all the segments of around 2.2 %. There is no significant change if the interpolation is of second degree. However, if cubic interpolation is used in this case it reduced to nearly 0.8 %. However, an exception is observed in first segment where the positional error in case of cubic interpolation has not shown any reduction in positional error. This large positional error in first segment can further be reduced by taking one additional accuracy point at one tenth of the first segment from zero. Using this method, the first segment positional error using cubic interpolation came down to the level comparable to succeeding segments. However, it is further observed that quadratic interpolation in the first segment has still further reduced positional error.

Overall, it is observed that cubic interpolation for determining the intermediate position of θ_1 and θ_2 is advantageous compare to linear and quadratic interpolation. This investigation needs to be further extended to optimize the interpolation and jerk at accuracy point. This can further be extended, if the velocity profile along the path is provided.

IOP Conf. Series: Materials Science and Engineering 1004 (2020) 012012 doi:10.1088/1757-899X/1004/1/012012

References

[1] Luh J Y S and Lin C S 1980 Optimum path planning for mechanical manipulators *Am. Soc. Mech. Eng.*, **104**, 142-151.

[2] Chen J and Chao L M 1987 Positioning error analysis for robot manipulators with all rotary joints *IEEE J. Robot. Autom.*, **3**(6), 539–545.

[3] Akima H 1970 A new method of interpolation and smooth curve fitting based on local procedures *J. ACM*, vol. **17**(4), 589–602.

[4] Keith L 1985 On-Line Polynomial Trajectories for Robot Manipulators *The international journal of robotic research*, **4**(2), 38-48.

[5] Rout B K and Mittal R K 2008 Parametric design optimization of 2-dof r-r planar manipulator-a design of experiment approach *Robot. Comput. Integr. Manuf.*, **24**(2), 239–248.

[6] Savsani P et al 2014 Comparative study of different metaheuristics for the trajectory planning of a robotic arm *IEEE Systems Journal*, 1–12.

[7] Gasparetto A et al 2015 Path planning and trajectory planning algorithms: a general overview *Motion and Operation Planning of Robotic Systems,* **1**, 3-27.

[8] Ichida K et al 1977 Curve fitting by a one-pass method with a piecewise cubic polynomial *ACM Trans. Math. Softw.*, **3**(2), 164–174.

[9] Ellis T R and McLain D H 1977 Algorithm 514: a new method of cubic curve fitting using local data *ACM Trans. Math. Softw.*, **3**(2), 175–179.

[10] Mittal R K and Nagrath I J 2003 Robotics and control *Tata McGraw Hill Publication,* New Delhi.

IOP Conf. Series: Materials Science and Engineering 1004 (2020) 012013 doi:10.1088/1757-899X/1004/1/012013

Geometric optimization of planar membrane reflector used in gossamer space structure

S. D. Shinde[1], M. Soni[2], S. H. Upadhyay[3*], S. Sakhare[4] and K. S. Singh[5]

[1,2,3]Smart Material and Structures Laboratory, Mechanical and Industrial Engineering Department, Indian Institute of Technology (IIT), Roorkee, 247667, India.
[4,5]Space Applications Centre-Indian Space Research Organisation, Ahmedabad, 380015, India.
*Email: sanjay.upadhyay@me.iitr.ac.in

Abstract. In recent trends membrane reflector has increased its applications in space missions due its significant advantage such as light in weight, higher folding and packaging efficiency, ease of deployment, low on-Board volume requirement. Geometry of the membrane reflector has significant influence on its the stress distribution and area of deformation, subsequently determining the stiffness and geometrical parameters of its supporting structures. The present study uses a Taguchi optimization method of orthogonal array for investigating the membrane geometry parameter and its effect on membrane reflector.

1. Introduction

Light in weight, higher folding and packaging efficiency, ease of deployment which makes gossamer structures best suit for the space applications. Membrane Reflector antenna is one the type of gossamer structure which used for efficiently distributing the electromagnetic waves in predefined manner. Such structures can be potentially used in near term space missions, as it giving an additional benefit such as reduction in total system mass and deployment complexity also reducing the size of launch vehicle [1]. As the reflector aperture area has direct relation with the gain from the antenna, hence the size and geometric specification of the reflector is important contributing factor in designing of large size space reflector antenna [2]. Hedgepeth et al. discussed the design requirements of large structures which are deployed, erected, assembled, or fabricated in space with high stability and accuracy [3].
Geometry and boundary support of membrane structure has significant influence on the material stresses, deformation areas and RF gain of antenna performance. M.J. Coleman et al. has investigated the influence of boundary support on performance characteristic of inflatable parabolic reflector antenna [4]. Circular membrane reflector with inflatable supporting structure has been developed by D P Cadogan et al. Polyamide is used as a membrane material for investigating the antenna characteristics with 16 catenary points [5].

Flow chart of the optimization process for different cutting pattern for membrane structure given by Kim J.Y et. al. Actual stresses and design stresses are used as key parameter for obtaining optimum cutting pattern [6]. Liu Z. et. al discussed the planar rectangular as well as circular membrane reflector antennas developed by Jet Propulsion library (JPL). Synthetic aperture radar antenna with a three-layer membrane reflector was developed by Canadian Space agency in 2007. Cutting pattern analysis for inflatable parabolic antenna with planar gores was carried out by Xu Y at. al. Shape analysis of inflatable parabolic reflector can be carry out by membrane theory and elastic mechanics [7-9].
Many researchers have used a planar membrane reflector with inflatable torus (or supporting) structure but detailed geometry optimization study was not done till date. The geometry and number of attachment

point of the membrane reflector has significant influence on the membrane supporting structure in terms of stiffness and its geometric parameters. The presented study uses a square membrane reflector with four and eight holding points with variation in geometry associated with changes in holding point is analyzed for material stresses and deformation area. The analysis is made under the assumption of deformation at the loading point should be less than 0.01 mm. The present work uses a Taguchi method of orthogonal array for finding the optimum geometric shape of square membrane reflector.

2. Methodology

2.1. Geometry and Material Specification

A square membrane of 500 mm side made of Kapton material has been selected for the investigation due to its inherent property of best suit to the space environmental conditions, Material properties of Kapton material are shown in Table 1. By keeping the reflector effective area for of 500 mm × 500 mm, outer geometry of the membrane is made of shape to facilitate ease of holding with its supporting structure and provide uniform stress distribution in the membrane reflector. The outer geometry of the membrane is highly dependent on the number of holding points. The holding points are increase from 4 to 8 and variation in geometry parameter is made to finding out material stresses, area of deformation and mass of membrane reflector. Currently membrane reflector with 4 and 8 holding points are presented in this manuscript. For four holding point, the geometry parameters are i. Pin length (P1), ii. Radius of Curvature (P2), iii. Corner Distance (P3) as shown in Figure 1. The one more parameter in eight holding points is added which is periphery distance(P4) as shown in Figure 2.

Table 1: Material property of Kapton material

Property	Kapton VN, 25 μm
Density (Kg/m^3)	1420
Poisson's ratio	0.34
Young's Modulus (MPa)	2500
Yield point at 3% (MPa)	69
Ultimate Elongation (%)	72

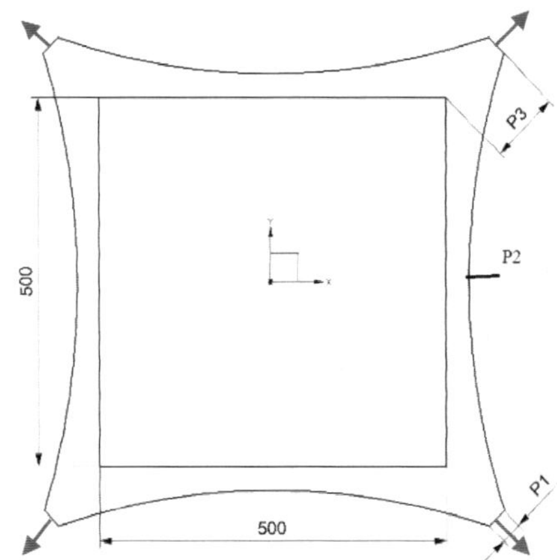

Figure 1: Membrane reflector with four holding point configuration

IOP Conf. Series: Materials Science and Engineering 1004 (2020) 012013 doi:10.1088/1757-899X/1004/1/012013

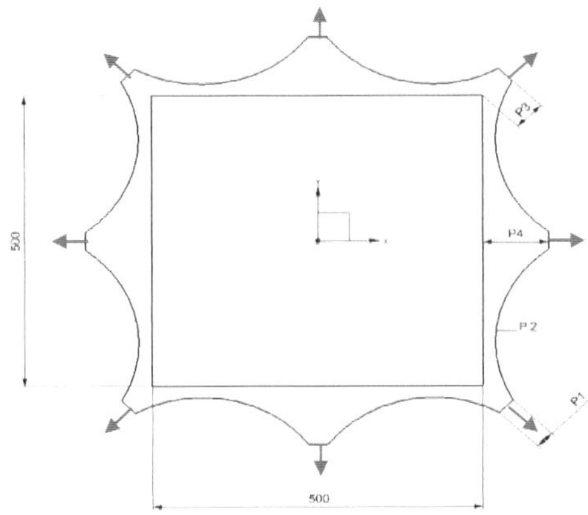

Figure 2: Membrane reflector with eight holding point configuration

2.2. Taguchi Orthogonal Array

Taguchi method of orthogonal array is one of the widely used as optimization technique which gives the information about conducting minimum number of experiment. This method is derived from the full factorial design approach. Taguchi Orthogonal Array gives the parameter variation table with minimum number of possible combination affecting the performance of the system [10]. For four holding point the three parameter are checked at 5 levels which gives the 25 experiment combination as per Taguchi orthogonal array method as shown in Table 2, and Table 4. Similarly, for eight holding point the four parameters are analyzed at four levels leads to 16 experiment combinations as shown in Table 3, and Table 5. The values of parameters and levels are selected for minimum material condition for the outer geometry of the membrane ultimately reducing the mass of membrane structure. The tension forces applied at the holding points are increased gradually till the 0.01mm of deformation and stress results are recorded.

Table 2: Parameter and Level for four holding point configuration

Four Holding Points			
Level	Pin length P1	Radius of Arc P2	Conner Distance P3
Level 1	16	800	55
Level 2	18	900	65
Level 3	20	1000	75
Level 4	22	1100	85
Level 5	24	1200	95

Table 3: Parameter and Level for eight holding point configuration

Eight Holding Points				
Level	Pin length P1	Radius of Arc P2	Conner Distance P3	Periphery Distance P4
Level 1	18	225	60	50
Level 2	20	250	70	60
Level 3	22	275	80	70
Level 4	24	300	90	80

Table 4: Taguchi array for four holding points

Four Holding Points			
Run	P1	P2	P3
1	16	800	55
2	16	900	65
3	16	1000	75
4	16	1100	85
5	16	1200	95
6	18	800	65
7	18	900	75
8	18	1000	85
9	18	1100	95
10	18	1200	55
11	20	800	75
12	20	900	85
13	20	1000	95
14	20	1100	55
15	20	1200	65
16	22	800	85
17	22	900	95
18	22	1000	55
19	22	1100	65
20	22	1200	75
21	24	800	95
22	24	900	55
23	24	1000	65
24	24	1100	75
25	24	1200	85

Table 5: Taguchi array for eight holding points

Eight Holding Points				
Run	P1	P2	P3	P4
1	18	225	60	50
2	18	250	70	60
3	18	275	80	70
4	18	300	90	80
5	20	225	60	60
6	20	250	70	70
7	20	275	80	80
8	20	300	90	50
9	22	225	60	70
10	22	250	70	80
11	22	275	80	60
12	22	300	90	50
13	24	225	60	80
14	24	250	70	50
15	24	275	80	60
16	24	300	90	70

Above geometrical parameter was investigated for combination given in Taguchi Orthogonal Array for stress distribution and mass criteria with the targeted or required deformation of 0.01 mm at the periphery of the membrane is shown in Table 4, and Table 5. The force values for required deformation are also found in each run of simulation in Abaqus Ver.2016 software [11].

3. Result and Discussion

For each run of the Taguchi Orthogonal Array, mass properties along with stress distribution and deformation area are investigated. The preliminary result shows that the stress distribution become more evenly and area of deformation are also reduced as the holding points are changed from four to eight. The graph for stress distribution along with the diagonal path of square membrane for each run of Taguchi orthogonal Array is plotted and it is observed that for four holding point the maximum and minimum stresses at periphery of 500 mm × 500 mm membrane was 0.2065 MPa and 0.16886 MPa for Run number 20 and 21 respectively. The minimum mass is observed for Run 11 of 10.3 gm with stress level of 0.2029 MPa as shown in Figure 4.

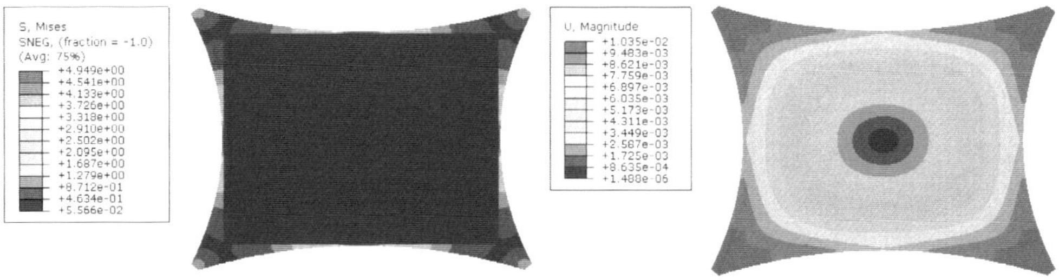

(a) Stress Distribution plot (b) Deformation plot

Figure 3: Numerical simulation result for four holding points

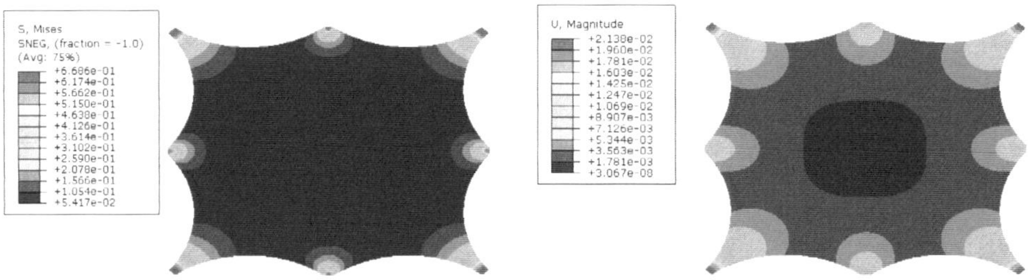

(c) Stress Distribution plot (d) Deformation plot

Figure 4: Numerical simulation result for eight holding points

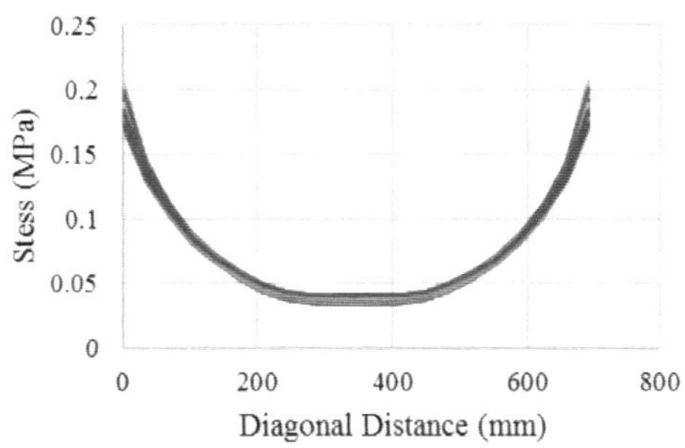

Figure 5: Variation of stresses along diagonal of RF area for four holding points

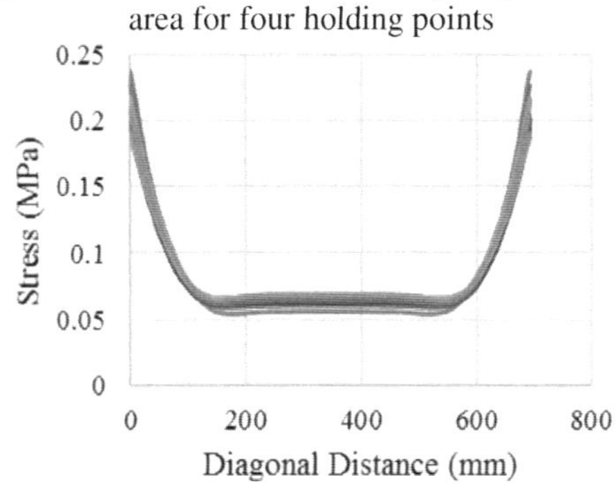

Figure 6: Variation of stresses along diagonal of RF area for eight holding points

Similarly, for the eight holding points the maximum and minimum stresses at the periphery of membrane 500 mm × 500 mm was 0.2367 MPa and 0.1889 MPa for the Run number 1 and 15 respectively. The minimum mass of 10.2 gm is also observed at Run 1. From the preliminary results it is observed that geometry has significant influence on mass property, stress distribution and force required for holding the membrane reflector. The deformation areas also reduced with increased in holding point. The force required for minimum mass array run geometry combination for membrane is found to be 0.48 N and 0.32 N respectively for four and eight holding points. These forces are the key factor for determining the stiffness and geometric parameters of the support structure.

4. Conclusion

With use of Taguchi orthogonal array parameter optimization technique, the geometry optimization was done for four and eight holding point of square membrane reflector. The values of stresses and forces required for targeted deformation was analyzed and it is observed that geometry has significance influence on these parameters. Further study with increased holding points in subsequent steps are needed for finding the best suit for geometry configuration and holding point needed.

Acknowledgments

The authors thankfully acknowledge the financial support provided by Space Applications Centre (SAC/ISRO) under respond project, Government of India. (Project Grant Number ISR-1248-MID)

References

[1] Chmielewski A B 2001 Overview of Gossamer Stuctrues *Gossamer Spacecraft:Membrane and Inflatable stuctures Technology for Space Applications*, edited by Paul Zarchan, Progress in Astronautics and Aeronautics AIAA Virginia 1-33

[2] Romanofsky R R, Lambert, Welch B W, Bibyk 2006 The Potential for Gossamer Deployable Antenna System in Ka-Band Exploration and science Communications Architectures *Twelth Ka and BroadbandCommunications Conference,* Italy

[3] Hedgepeth J 1891 Critical requirements for the design of large space *structures Conference on Large Space Platforms: Toward Permanent Manned Occupancy in Space* 443

[4] Coleman M J and Baginski F 2012 Effect of Boundary Support and Reflector Diamension on Inflatable Parabolic Antenna Performance *Journal of Spacecraft and Rockets,* **49,** 5

[5] Cadogan D P, Lin J K, Grahne M S 1999 The development of inflatable space radar reflectarrays *Proceedings of the 40th AIAA/ ASME/ ASCE/AHS/ASC Structures, Structural Dynamics, and Materials Conference and Exhibit* Saint Louis Missouri

[6] Kim J Y and Lee J B 2012 A new technique for optimum cutting pattern generation of membrane structures *Engineering Structures,* **24,** 745–756

[7]Liu Z, Qiu H, Li X, Yang S 2017 Review of Large Spacecraft Deployable Membrane Antenna Structures *Chinese Journal of Mechanical Engineering* 1447–1459

[8]Xu Y, Guan F, 2012 Structure design and mechanical measurement of inflatable antenna *Acta Astronautica,* 13–25

[9]Lee, K and Lee, S W 2006 Analysis of gossamer structures using assumed-strain solid-shell finite elements *Journal of spacecraft and rockets,* **43,** 1301-1307

[10] Cimbala J M 2014 Taguchi Orthogonal Arrays Penn State University

[11] Abaqus, Simulia Ver. 2016 Johnston Rhode Island United States

IOP Conf. Series: Materials Science and Engineering 1004 (2020) 012014 doi:10.1088/1757-899X/1004/1/012014

Multi Objective Optimization of Nano-composite Non-asbestos Friction Lining Material using Grey Relation Analysis (GRA)

Dinesh Shinde[1, 3*], Mukesh Bulsara[2, 3] and K N Mistry[3]

[1]SVKM's NMIMS MPSTME Shirpur, Dist-Dhule, MH, India.
[2]G H Patel College of Engineering & Technology, V.V. Nagar, Anand, Guj, India.
[3]Gujarat Technological University, Ahmedabad, Guj, India
[*]Email: dinesh.shinde61@gmail.com

Abstract. Friction lining materials should possess high coefficient of friction at the same time should have high wear resistance. Moreover, it should be stable at elevated temperature encountered due to sliding friction. One has to think of many functional requirements in order to develop a friction material,, which gives motivation to optimize the friction materials for multiple objectives. In this work, three variant (CSP0, CSP15 and CSP30) of friction lining materials with nano-coconut shell powder are developed and subjected chase friction testing. Specimen preparation is done at Universal brakes Jalgaon (MS, India) and Chase friction test is conducted at Indian Friction Material Engineering Company, Delhi (India). The results of the test are then used in multi-objective decision making for the selection using Grey Relation Analysis (GRA). It is applied based on the temperature sensitivity performance of Coefficient of friction (CoF) of the specimens, through six different criterions in parallel. It is observed that CSP30 has ranked 1 according to GRA with six criterions.

1. Introduction

Friction lining materials are the category of material which should wear progressively giving desired frictional effect and also sustain their properties at elevated temperature. This functional requirement motivates to develop the material with more devotion and determination. These are the composite materials with five major components, i.e. binder, reinforcing fiber, filler, friction modifier and additives. In each of the category, more than one material can be used to improve overall properties of the friction material. Till 90's, the friction lining materials consisted of asbestos as a major constituent, due to its dominant mechanical and physical properties. Environmental issues related to asbestos led to development of non-asbestos friction lining material [1]. Researchers have tried metallic, semi-metallic and non-metallic material to replace asbestos [2-12]. Again, due to ecological issues related to copper, in this decade many countries have banned the use of copper as friction lining materials, which further led to development of copper free friction lining material [13-15].
Nowadays, nano-materials have attracted the scientist to use them in composite material, due to exhibition of desired properties in the resulting material. The friction lining material with nano-material as a constituent shows enhanced tribological and mechanical properties [16]. Nano-metallic compositions in the friction lining material with improved properties are being used and reported by some researchers [17-18]. But, the friction lining material with non-metallic, eco-friendly compositions are still under development.
The sliding speed, braking force and sliding distance are the performance parameters of the automotive brakes. These performance characteristics of the brake depends upon the tribological

properties of the friction lining material, which slides over the drum or disc surface to provide the braking. Qi et al. (2007) had investigated the interface temperature between the disc and friction lining pad considering these parameters [19]. Ficici et al. (2011, 2014)[20, 21] and Kumar et al. (2013)[22] considered material compositions with these parameters to optimize the tribological behavior using Taguchi method, whereas Kim et al. (2003)[23] optimized the manufacturing parameters of the friction lining material with these operating parameters of the brake. Researchers commonly use Taguchi method to optimize the performance of friction lining material [24-25].

Need of Today's growing world is to optimize the output with multiple objective for minimum resources, which stimulated the evolution of multi-objective optimization. Grey Relational Analysis (GRA) integrated with Taguchi Design is one of those kinds. Manikandan et al. (2017) had used GRA method to optimize the electrochemical drilling of Inconel 625 for higher material removal and lower surface roughness [26]. Mozammel et al. (2016) optimized the process variables of high-pressure turning of Ti-6Al-4V with coolant for minimum cutting forces and surface roughness, using sole objective and multiple objective approach and compare them with each other [27]. Farhad et al. (2012) optimized the turning process of 50.2 steel for maximum material removal rate and lower surface roughness [28]. Milling parameters of titanium alloy rotor disc with blades channels has been optimized for higher amount of materials removed and life of the cutter, for minimum thickness of residual stress layer [29]. Researchers have applied GRA optimization method to different field of interest [30, 31].

Despite of enormous research in the field of non-asbestos friction lining material, the application of multi-objective optimization to nano-composite friction lining materials for higher coefficient and lower wear is untouched. An attempt to elaborate this, through experimental and statistical techniques is made here, through development of a non-asbestos friction lining material using a nano-coconut shell powder as a filler keeping other compositions fixed and subjecting it to an experimentation design using GRA multi-objective optimization technique.

2. Materials and Methods

This work aims at analysing automotive drum brake using finite element analysis, for nodal deformation of the friction liner surface in contact with drum and validating the outcomes using a mean height measurement of the nodes respective to the surface of the friction liner after experimentation with similar performance parameters. A standard size our wheel automotive drum brake is modeled in software package Solidworks 17.0 available at CAD/CAE lab (Mechanical Engineering Department, MPSTME SVKM's NMIMS Shirpur, Dist-Dhule, MS), with shoe and friction liner shown in figure 1. Considering the symmetrical geometry of the drum brake, a half model is taken.

In the present study, two friction lining materials to be used in automotive drum brake, are developed varying the nano-coconut shell powder as a filler keeping remaining composition same. Coconut shells are collected, cleaned, crushed and sieved manually up to required size. The sieved powder is subjected to particle size measurement using Malvern Zetasizer. Pdl is the polydispersity which describes the degree of non-uniformity of a size distribution of particles in a powder. It is also called as heterogeneity index, which in our case is 0.618. The compositions shown in table 1, is selected with reference to literature [32-41].

Table 1: Composition of the materials

Sr. No.	Ingredients		Weight percentage		
			CSP0	CSP15	CSP30
1	Binder (Phenolic resin)		20	20	20
2	Reinforcement (Glass fiber and Steel wool)		15	15	15
3	Filler	Coconut Shell Powder (CSP)	0	15	30
		Calcium Carbonate ($CaCo_3$)	35	15	0

5	Friction Modifier (Al_2O_3 , Friction dust and Zr_2SO_4)	10	15	15
6	Solid Lubricant (Graphite powder)	5	5	5
7	Additives (NBR, Aramid pulp)	15	15	15

With these compositions, drum brake liners are manufactured at Universal Brakes Ltd, Jalgaon (MS-India). The ingredients are mixed thoroughly using mixer. The mixture is preformed and molded into slab of required shape but of larger size. The friction material slab is then put into electric furnace for sintering. The grounded piece of linings obtained from the slab are to be bonded over the brake shoe, which is accomplished through process of bonding and finally a brake liner with shoe is made ready. Chase friction testing is performed for tribological analysis. Fade & recovery are the phenomenon used to demonstrate the actual operating conditions of the brakes, which are utilized here. Table 2 shows the performance parameters of the test.

Table 2: Temperature sensitivity test circumstances

Operation cycle	Conditions
First fade	Starting temperature= 91^{0}C, Ending temperature= 289^{0}C, Sliding speed (rpm)= 400, applied pressure= 1 MPa, times of brakes applied in a cycle=8, gap between two brakes is 10 s, air blower=off.
First recovery	Starting temperature= 91^{0}C, Ending temperature= 289^{0}C, Sliding speed (rpm)= 400, applied pressure= 1 MPa, times of brakes applied in a cycle=5, and gap between two brakes is 10 s, air blower=on.
Second recovery	Starting temperature = 93^{0}C, ending temperature= 345^{0}C, sliding speed (rpm)=400, applied pressure= 1 MPa, times of brakes applied in a cycle =10, gap between two brakes is 10 s, air blower=off.
Second recovery	Starting temperature= 317^{0}C, ending temperature= 93^{0}C, sliding speed (rpm) =400, applied pressure= 1 MPa, times of brakes applied in a cycle =5, gap between two brakes is 10 s, air blower=on.

2.1 Grey Relational Analysis (GRA)

The grey relational analysis is used in this work considering seven criterions listed in table 3 [33-36]. It is a multi-criterion decision making methodology, which is utilised mostly to assess the raw data without any association during the operations. GRA is based on following steps: (1) recognition of evaluating criterions, (2) preparation of operations matrix, (3) preparation of normalized operations matrix, (4) evaluation of grey relation coefficient, and (5) evaluation of grey relation degree.

Step-1: Evaluation criterions are selected from performance parameters of the process under study. There may be n-number of criterions.

Step-2: The operations matrix is generated through criterions and possible ways to solve the problem. If solution ways are A and number of criterions are B then the operations matrix having an order of A × B is to be used.

Step-3: The operations matrix are then normalized between 0 and 1. Each value in the normalised operations matrix is determined using following relations:

Larger the better,

$$X'_{ij} = \frac{X_{ij}-min\{X_{ij}\}}{max\{X_{ij}\}-min\{X_{ij}\}} \quad (1)$$

Smaller the better,

$$X'_{ij} = \frac{min\{X_{ij}\} - X_{ij}}{max\{X_{ij}\}-min\{X_{ij}\}} \quad (2)$$

Step-4: Grey relation coefficient for each term is evaluated using relation:

$$\Psi = \frac{\Delta_{min}+\varepsilon\Delta_{max}}{\Delta_{0i}+\varepsilon\Delta_{max}} \quad (3)$$

Step-4: A grey correlation degree is determined using following relation:

$$\Phi = \frac{1}{N}\sum_{j=1}^{N}[\Psi] \quad (4)$$

Table 3: Details of the criterions

Criterion	Description
Crit-1 Coefficient of friction (μ) Higher-the-better	It is the mean CoF during all the test conditions.
Crit-2 Wear (gm) Lower-the-better	It is interminable material removal from the surface specimen.
Crit-3 Friction fade % Lower-the-better	Friction fade % = $\frac{\mu-\mu_F}{\mu} \times 100$, μ_F is the upper terminal value of CoF for the fade cycle.
Crit-4 Friction recovery % higher-the-better	Friction recovery % = $\frac{\mu_R}{\mu}\times 100$, μ_R is upper terminal value for CoF of the recovery cycle.
Crit-5 Stability coefficient Higher-the-better	It is the ratio between CoF and the upper terminal value of CoF i.e. $\frac{\mu}{\mu_{max}}$. It revealed how stable system response is.
Crit-6 Variability coefficient Lower-the-better	It is the ratio of top most value and lowest CoF, i.e. $\frac{\mu_{min}}{\mu_{max}}$. It shows volatality in the system response.
Crit-7 Friction fluctuation Lower-the-better	It is the variation in the top most value and lowest value of CoF during actual operation i.e. $\mu_{max} - \mu_{min}$.

3. Results and Discussions

Figure 1, 2 & 3 shows fade & recovery plots of CSP0, CSP15 and CSP30 respectively. First fade cycle starts from 910C and ends at 2910C, during this cycle CoF first increases, then decreases for all the three materials. First recovery starts when the fade ends and remains till 910C. Preferably second fade and recovery are taken to be of larger range of temperatures.

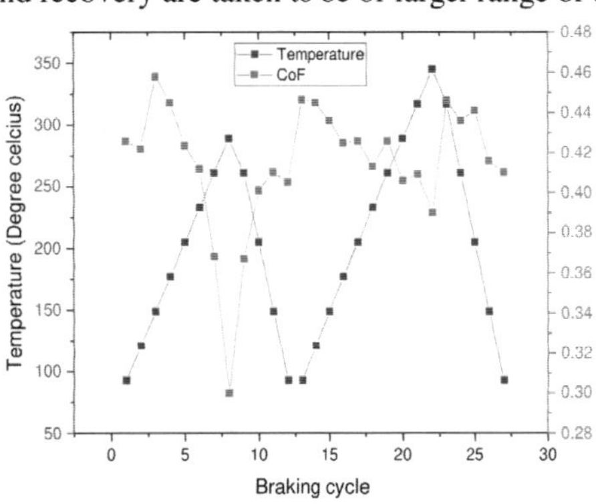

Figure 1: Fade & recovery plot of CSP0.

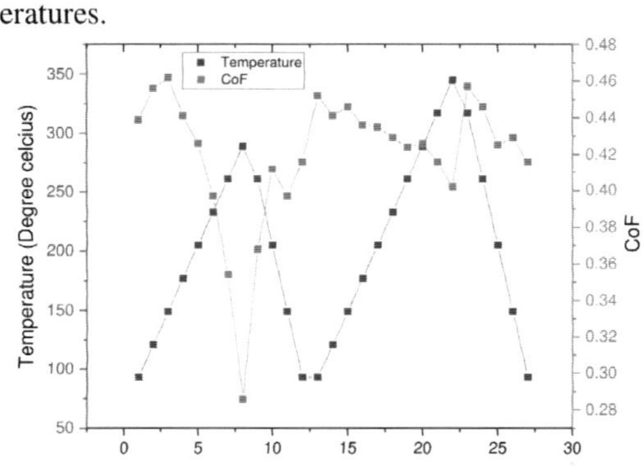

Figure 2: Fade & recovery plot of CSP15.

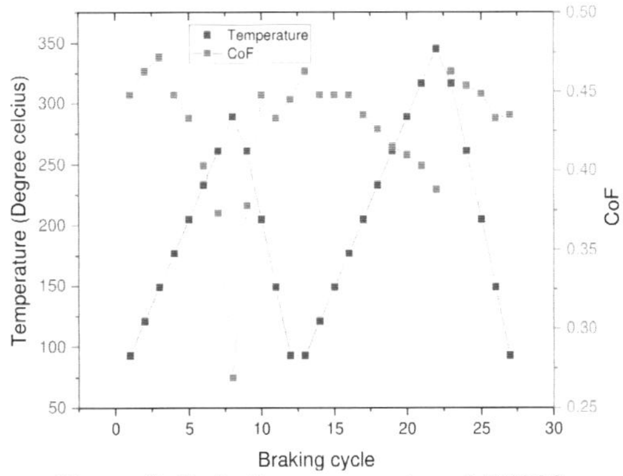

Figure 3: Fade & recovery plot of CSP30.

Table 4 shows normal decision matrix with reference sequence which shows average values of the terms used in defining the criteria, taken from the fade and recovery plots. Moreover, they are the values obtain from experimental results. Maximum and minimum values of the criterions are used for further calculations. Table 5 is the difference matrix, which is obtain through their performance factuality i.e larger-the- better or smaller-the-better. Table 6 shows the grey relation coefficient obtain using equation (3). Gray relation grades are shown in table 7. Grey relation degree is obtained through equation (4).

Table 4: The normal resolution matrix

Formulation	CSP0	CSP15	CSP30
Crit-1	0.400	0.411	0.438
Crit-2	10.260	9.870	9.450
Crit-3	23.576	24.579	29.604
Crit-4	9.280	10.255	11.052
Crit-5	94.787	95.426	95.842
Crit-6	77.293	75.758	75.847
Crit-7	0.104	0.112	0.114

Table 5: The difference matrix

Formulation	CSP0	CSP15	CSP30
Crit-1	0.00	0.29	1.00
Crit-2	-1.00	-0.52	0.00
Crit-3	0.00	-0.17	-1.00
Crit-4	9.28	10.26	11.05
Crit-5	94.79	95.43	95.84
Crit-6	-1.00	0.00	-0.06
Crit-7	0.00	-0.80	-1.00

Table 6: Grey relation coefficient

Formulation	CSP0	CSP15	CSP30
Crit-1	0.00	0.286321	1
Crit-2	2	1.518519	1
Crit-3	1	1.166348	2
Crit-4	-8.279	-9.255	-10.05
Crit-5	-93.78	-94.426	-94.84
Crit-6	2	1	1.0585
Crit-7	1	1.8	2

Table 7: Grey relation grade

Formulation	CSP0	CSP15	CSP30
Crit-1	1.0000	0.6359	0.3333
Crit-2	0.2000	0.2477	0.3333
Crit-3	0.3333	0.3001	0.2000
Crit-4	-0.0643	-0.0571	-0.0523
Crit-5	-0.0054	-0.0053	-0.0053
Crit-6	0.2000	0.3333	0.3208
Crit-7	0.3333	0.2174	0.2000

Table 8: ranking of formulations based on GRA

Formulation	Grey relation degree	Ranking
CSP30	0.285	1
CSP15	0.239	2
CSP0	0.190	3

Figure 4 shows wear of the three materials after test runs calculated considering mass before the and mass after the test. It is observe that CSP30 gives least wear among the materials, which ultimately confirm an important aspect of friction materials selection.

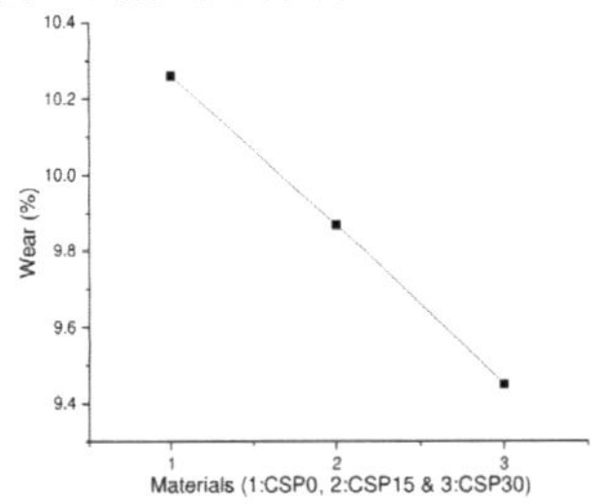

Figure 4: Wear of the materials.

4. Conclusions

Development and multi-criterion optimization of friction materials with nano-coconut shell powder led to following conclusions:

1. Chase friction testing is an important tribological testing methodology, which provides the crucial behavior of the friction material to perform in higher operating parameters.
2. The recovery performance remains highest for CSP0, which gets mitigated with higher fade performance.
3. GRA methodology is proved to be useful tool in design of friction materials formulation. Assessment of the friction materials can be done using GRA. Results of GRA have ranked the friction material CSP30 as the best, CSP15 as average and CSP0 as poor amongst the three. Higher percentage of CSP proves to be best for all the criterions.
4. Tribological and mechanical properties of CSP30 is found to be superior.

Acknowledgement

Authors acknowledged the Universal Brakes Ltd Jalgaon for help in friction materials formulation, Indian Friction Material Engineering Company- Delhi (India) for Chase testing of the friction materials.

References

[1] D Chan and G W Stachowiak 2004 *Proceedings of Institution of Mechanical Engineers Part D: J. Automobile Engineering* **218,** 953
[2] Min Hyung Cho *et al* 2005 *Wear* **258**1,682.
[3] Peter Filip *et al* 2002 *Wear* **252,** 189.
[4] Piyush Chandra Verma *et al* 2016 *Wear* **346-347,** 56.

[5] J. Abutu *et al* 2019 *SN Applied Sciences* **1,** 82.
[6] I. Mutlu *et al*, 2006 *Tribology International* **39,** 317.
[7] D.C. Jara and H. Jang 2019 *Wear* **05,** 11.
[8] M. W. Shin *et al* 2015 *Tribology Letters* **58,** 10.
[9] Samrat Mohanty and Y.P. Chugh 2007 *Tribology International* **40,** 1217.
[10] Vlastimil Mateˇjka *et al* 2013 *Materials and Design* **51,** 847.
[11] Leonardi M. *et al* 2017 *Tribology International* **10,** 37.
[12] Vishal Mahale *et al* 2019 *Wear* **02,** 19
[13] V. Mahale *et al* 2019 *Tribology International* **03,** 49.
[14] Cinzia Menapace *et al* 2017 *Materials and Design* **115,** 287.
[15] Jayashree Bijwe *et al* 2012 *Wear* **296,** 693.
[16] M. Amirjan 2018 *Tribology International* **10,** 40.
[17] M. Baklouti *et al* 2015 *Wea* **330-331,** 507.
[18] K. K. Ikpambese *et al* 2014 *Journal of King Saud University – Engineering Sciences* **102,** 11035.
[19] Zhanhong Wang *et al* 2016 *Materials and Design* **90,** 76.
[20] Shinde Dinesh *et al* 2019 *Proceedings of the Institution of Mechanical Engineers, Part-J:Journal of Engineering Tribology* **243(6).**
[21] Jayashree Bijwe *et al* 2008 *Wear* **265,** 699.
[22] Ferit Ficici *et al* 2014 *Journal of Brazillian Society of Mechanical Sciences and Engineering* **36,** 653.
[23] Mukesh Kumar and Jayashree Bijwe 2013 *Wear* **303,** 569.
[24] Seong Jin Kim *et al* 2003 *Journal of Materials Processing Technology* **136,** 202.
[25] Tej Singh *et al* 2015 *Wear* **324-325,** 10.
[26] Dragan Aleksendric and Adolfo Senatore 2012 *Journal of Composite Materials* **46(22),** 2777.
[27] N. Manikandan *et al* 2017 *Engineering Science and Technology, an International Journal* **20,** 662.
[28] Mozammel Mia *et al* & 2016 *International Journal of Advanced Manufacturing Technology* **170,** 9372.
[29] Farhad Kolahan *et al* 2012 *Applied Mechanics and Materials* **110-116,** 2926.
[30] Zhishan Li and Yaoyao 2019 *Symmetry* **11(173),** 11020173.
[31] Shreemoy Kumar Nayak *et al* 2014 *Procedia Materials Science* **6,** 701.
[32] A. Noorul Haq *et al* 2008 *International Journal of Advanced Manufacturing Technology* **37,** 0981-4.
[33] T. Singh *et al* 2015 *Materials and Design* **10,** 45.
[34] D. Aleksendric and C. Duboka 2006 *Wear* **261,** 269.
[35] D. Aleksendric and C. Duboka 2007 *Wear* **262,** 778.
[36] B. K. Satapathy *et al* 2006 *Journal of Composite Materials* **40(6),** 483.

Design Improvements in Conventional Drilling Machine to Control Thermal Necrosis during Orthopaedic Surgeries

R V Dahibhate[1*], S B Jaju[2] and R I Sarode[3]

[1]Research Scholar, G.H.Raisoni College of Engineering, Nagpur, Maharashtra, India
[2]Professor, G.H.Raisoni College of Engineering, Nagpur, Maharashtra, India
[3]Professor, Dr. Ulhas Patil Medical College and Hospital, Jalgaon, Maharashtra, India
[*]E-mail: dahibhaterv@gmail.com

Abstract: Bone drilling operation is performed to join fractured bone tissues during physical trauma. Natural bone growth and remodelling of fractured bone fragments is possible by immobilisation of affected bone. Immobilisation is achieved by use of screws and plates with drilling facilities. Similarly internal fixation of screws [1] during intramedullary nailing often involves bone drilling. The heat generation model [2] has proved that, generation and accumulation of heat above temperature level of 47^0c for one minute during drilling is the alarming signal for thermal necrosis which causes irreversible bone damage. To study bone drilling in concern with thermal necrosis, threshold temperature has to be the prime concern and accordingly drilling parameters and specifications are to be selected. In this study, drilling process is conducted on a sheep rib bone. Optimization of drilling parameters such as drilling speed, feed rate and drill specification like drill tool diameter is suggested using Taguchi and ANOVA method. Results indicate drill diameter is most influencing factor contributing 90.71% for heat generation, followed by feed rate 5.69% and drill speed 2.22%. Based on discussions and conclusions a drill machine with attachments of infrared thermometer, speed sensor and camera is designed and fabricated. This machine alarms surgeon when temperature level at drill site increases above 47^0c, which is a threshold temperature of thermal necrosis.
Keywords: Bone drilling, thermal necrosis, threshold temperature, optimization, infrared thermometer

1. Introduction

The bone is physiologically active and reactive tissue and explained chemically as Ca10(PO4)6(OH)2. The axial and appendicular skeleton structure forms the supporting frame for soft tissues, organs and muscles attachments of an animal body. Bones take the substantial load during body movement, lifting weights and running. Bones are calcified connective tissue, comprising of minerals, collagenous protein, lipids and water.

Drilling in bone tissue is a better option for insertion of fixations. Depending on the severity of accidents and trauma situations the numbers of drills require may be in multiples. The bone drilling is a precision work, from mechanical engineering aspect [3] as it involves handling of drilling machine with controlling of spindle speed, manual feed rate, depth of cut, drill tool geometry, chip flow rate and from therapeutic aspect the bone cell thermal damage (thermal necrosis) [4] and patient early rehabilitation are prime considerations and so this precision task demand knowledge, skill and experience of surgeons. In previous research work it is clear that if the temperature at drilling location reaches more than 470c and remains the same for one minute, it could damage the cells permanently and so considered as threshold temperature. During interactive sessions with medicinal experts, currently to control heat generation a number of drilling techniques along with coolant supply, like push hard i.e. more feed rate or push-pull i.e. removing the drilling tool number of times from the drill hole are adopted. The causes of heat accumulation while bone drilling can be explained as; shear deformation across shear region [5], rubbing in between the drill bit [6] rake face, chip, friction among the tool flank face and recently formed work piece wall's [7] and so to have an accurate information about temperature level there is necessity to make improvements in the existing conventional drilling machine by the fitments like infrared thermometer and camera to alarm the surgeons about thermal necrosis.

2. Materials and Methods

The aim of research is, to make design improvements in conventional drilling machine and thereby avoid thermal insult in the course of orthopaedic drilling. For drilling sheep rib bone is selected because it is a suitable large-animal typical for biomedical research. Other factors are availability, ease of supervision and housing, animal price, and receiving to society as a research animal [8, 9and 10]. In this study, sheep rib bone acquired from a local butcher. First, all the soft flesh tissues were detached along the specimen surface by a surgical tool after that the bone specimen kept in buffer solution. To achieve a realistic environment of surgery and accuracy, while drilling the bone with conventional drilling machine sample is placed in a special container. The container is filled with saline water, with a provision of heating the normal saline solution and a thermostat to maintain a precise body temperature 37^{0}C.The average dimensions of specimen were thickness 2.5 mm and surface area 10mmX6mm, across which heat distribution is to be studied. The known bone properties [11] were as in table 1.

Table 1:Thermal properties of bone

property	Value
Thermal conductivity (W/mk)	0.54
Specific heat (J/(kgK))	1260
Density(kg/m3)	1800

To measure the temperature of bone at heat affected drilling site, infrared thermometer (specifications in Table No. 6) is used. Thermocouple's distant location from drilling site, thermal conductivity of bone material is the limitations that can be avoided by infrared thermometer. At every experiment cycle, drilling was performed and at the same time temperature was measured. To know the effect of three known variables drill tool diameter, drilling speed [12] and drilling feed rate on bone temperature, which is an unknown variable or response, experimentation is conducted. For experimentation purpose, three different drill tools [13] of diameter 2.5, 3.2 and 4.5 mm are used. Drill tool diameters are selected considering the standard drill diameters frequently used by surgeons. Tool geometry features helix angle 300, point angle 1180 and web thickness 0.9 mm. Three feed rate values 50.60 and 70 mm/min. and three drill speed ranges 1000, 1500 and 2000 rpm were selected. In this way, combination of three control factors [14] viz, spindle speed, feed rate and drill diameter, with three levels form 27 combinations [15]. 27 sheep rib bone samples were drilled.

Table 2: Drilling control factors and levels

Levels	Drill Diameter (mm)	Feed Rate (mm/min)	Spindle Speed (rpm)
Level 1	2.5	50	1000

IOP Conf. Series: Materials Science and Engineering 1004 (2020) 012015 doi:10.1088/1757-899X/1004/1/012015

Level 2	3.2	60	1200
Level 3	4.5	70	2000

Table 3: Temperature of bone sample at drilling site and Completed orthogonal array L_{27}

Experiment Number	Drill diameter (mm)	Feed rate (mm/min)	Drill speed (rpm)	Temperature (°c)	S/N ratio (dB)
1	2.5	50	1000	41.32	-32.3232
2	2.5	50	1500	41.01	-32.2578
3	2.5	50	2000	39.72	-31.9802
4	2.5	60	1000	40.91	-32.2366
5	2.5	60	1500	40.21	-32.0867
6	2.5	60	2000	38.9	-31.799
7	2.5	70	1000	40.1	-32.0629
8	2.5	70	1500	39.12	-31.848
9	2.5	70	2000	37.51	-31.4829
10	3.2	50	1000	45.93	-33.2419
11	3.2	50	1500	45.21	-33.1047
12	3.2	50	2000	44.56	-32.9789
13	3.2	60	1000	43.73	-32.8156
14	3.2	60	1500	43.29	-32.7278
15	3.2	60	2000	42.19	-32.5042
16	3.2	70	1000	43.21	-32.7117
17	3.2	70	1500	43.19	-32.7077
18	3.2	70	2000	42.61	-32.5902
19	4.5	50	1000	49.12	-33.8252
20	4.5	50	1500	49.31	-33.8587
21	4.5	50	2000	48.62	-33.7363
22	4.5	60	1000	48.19	-33.6591
23	4.5	60	1500	47.92	-33.6103
24	4.5	60	2000	47.69	-33.5685
25	4.5	70	1000	47.12	-33.4641
26	4.5	70	1500	47.19	-33.477
27	4.5	70	2000	46.72	-33.3901

To confirm the effect of temperature [16] on bone cells, drilled samples were checked by histopathology. For histopathology [17], paraffin method is preferred [18]. Only the drilled bone samples affected by thermal necrosis [19] i.e. samples with sufficient empty vacuoles were considered. In case of histopathological failure, the particular experiment was repeated with the same three drilling parameters assigned to that experiment number.

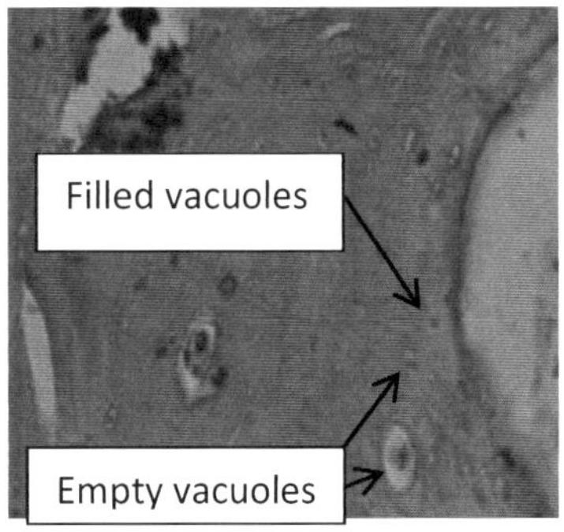

Figure 1: Histopathology of bone tissue

Using Minitab software Taguchi Analysis [20] is performed. As during drilling process the minimum temperature is preferred, to achieve optimum grouping of control factors, smaller the best (SB) quality characteristics is selected. The equation for Signal to noise ratio is,
$S/N = -10 \text{ Log } \Sigma (Yi^2/n)$, Where, Y_i is temperature response in that particular run and n is sample size.

3. Results

Table 4: Average S/N ratio of all factors at each level

Levels	Drill Diameter (mm)	Feed Rate (mm/min)	Spindle Speed (rpm)
Level 1	-32.01	-33.04	-32.82
Level 2	-32.82	-32.78	-32.87
Level 3	-33.62	-32.64	-32.77
Delta	1.61	0.40	0.10
Rank	1	2	3

Signal to noise is a quality indicator for the scattering around target values. High S/N ratio is indication of high signal value than random noise value. Average value for each control factor is calculated by using Minitab. Average value for level 1, drill diameter i.e. 2.5mm is calculated by taking the average of signal to noise ratio value related to 2.5mm i.e. experimentation number 1 to 9 from table 2. (-32.3232) + (-32.2578) + (-31.9802) + (-32.2366) + (-32.0867) + (-31.799) + (-32.0629) + (-31.848) + (-31. 48 29) / 9 = -32.0086 dB equals to -32.01dB, similarly average values of other control factors are calculated. Delta is calculated by subtracting the maximum and minimum values of average responses, for example Delta value for drill diameter is equals to (-32.01) - (-33.62) = 1.61.The Rank indicates the ranking of every delta value among the other ones, for example the delta value for drill diameter is 1.61, for feed rate [21] is 0.40, spindle speed is 0.10 and so rank of drill diameter is 1.Rank also indicates the influencing capacity of that factor.

Table 5: Percentage contribution of control factors using Anova method

Source	DF	Seq SS	Contribution (%)	Adj SS	Adj MS	F-Value	P-Value

Drill Diameter	2	296.85	90.71	296.85	148.42	659.93	0.00
Feed Rate	2	18.62	5.69	18.622	9.31	41.40	0.00
Spindle Speed	2	7.275	2.22	7.275	3.63	16.17	0.00
Error	20	4.498	1.37%	4.498	0.22		
Total	26	327.25	100%				

Increased samples quantity gives better observation base and so 27 response values are recorded. Total number of observations (n) are 27 and DF= (n-1), hence total of degree of freedom (DF) is 26.Error indicates the observations with each predictor having same value and as in the observation three control factors contributing to 27 readings, the error comes to 20. Sequential sums of squares (Seq SS) depend on the order the terms are entered into the model and are measures of variation for different components of the model, drill diameter data is entered in most sequentially and so drill diameter Seq.SS is 296.85.whereas spindle speed values are most disturbed and so its value is 7.275. The adjusted sum of squares (Adj SS) quantifies the amount of variation in the response data that is explained by each term in the model. A sufficiently large F-value indicates that the term or model is significant, drill diameter most significant with F-value of 659.93 and spindle speed the least with F-value 16.17.

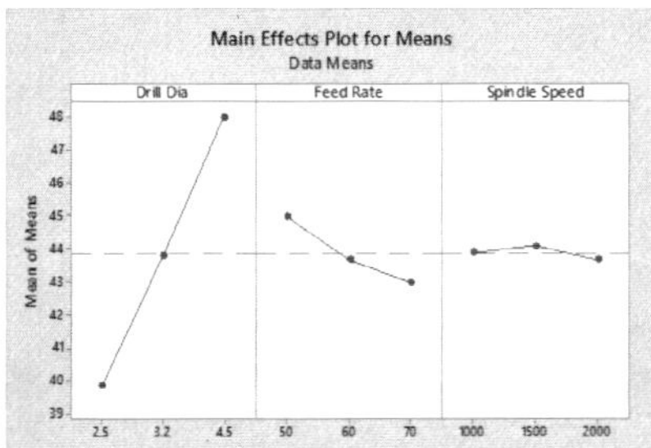

Figure 2: Main effects plot for mean of means

Response curve indicates graphical representation of change of performance characteristics with the change in drilling parameter level. Figure 2 shows the response graph for drill diameter, feed rate and spindle speed.

4. Improved Orthopaedic Drilling Machine

It is observed that if an orthopaedic drilling machine is equipped with some basic electronic devices like thermal sensors, that can keep inform the surgeon about temperature level at drilling site, in this way, thermal necrosis can be controlled. For this the temperature at drilling site should be monitored continuously and a signal should be sent to indicate if temperature is above 47^0c .To realize this, a non-contact, distantly temperature measuring Infrared thermometer is mounted on drilling machine.

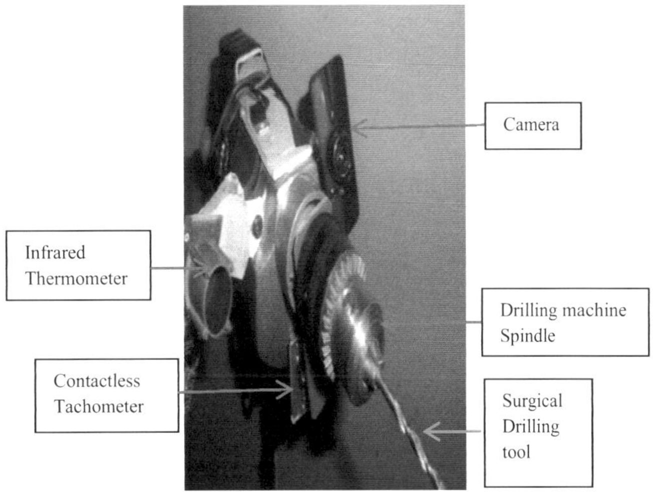

Figure 3: An orthopaedic drilling machine

4.1 Drill specifications

Table 6: Drill machine and infrared thermometer specifications

A)	Drill machine specifications			
	1	No load speed	1st gear	0-500 rpm
			2nd gear	0-1900rpm
	2	Torque	soft material	21 Nm
			Hard material	54 Nm
	3	Chuck capacity	Minimum	1.5 mm
			Maximum	13 mm
	4	Tool holder type	Three-jaw chuck	
	5	Number of torque settings	20+1	
	6	Battery type	Lithium-ion	
B)	Infrared thermometer specifications			
	1	Model		GY-906MLX90614
	2	Temperature Range		-20^{0}C to 320^{0}C
	3	Temperature Accuracy		±2^{0}c
	4	Repeatability		2%rdg
	5	Emissivity		0.95
	6	Spectral Response		07-18μ
	7	Response Time		500 msec
	8	Minimum Spot Size versus Distance ratio		6:1

9	Laser diode (aiming of infrared thermometer)	50 mW
10	Temperature display	NPN 7 channel transistor with light emitting diode
11	RPM sensor	
12	Camera	
13	Buzzer unit	

Three electronic circuits are fabricated: i] Main Controller Circuit: Orthopaedic drill machine and infrared thermometer, ii] Receiver Circuit: To receive the outputs about drilling site temperature, process the inputs and signal surgeons in terms of digital signal and alarming buzzer, that temperature is above threshold levels and the rectified measures like withdrawal of drilling tool, lowering of drilling speed or feed rate, increasing coolant supply can be adopted. iii] Charging circuit: For continuous and uninterrupted power supply Lithium-ion batteries with is used. The suggested drilling machine with continuous temperature monitoring [22] offers not only accurate control of heat but also time factor, otherwise temperature raise above 47^0c ,have to wait for 11 seconds. In this way before attaining temperature up to 47^0c, the corresponding signal received by infrared thermometer and buzzer alarms surgeon. The suggested drilling machine 1) avoids thermal necrosis by controlling the drilling speed, as from the two drilling parameters; speed and feed rate, only speed can be monitored and governed and feed rate is provided manually. Similarly with respect to bone dimensions drill diameter has to be selected without an alternative but here also speed variation can avoid necrosis2) reduce surgery time duration, as above attainment of47^0c, cooling waiting period is required, 3) offers hassle free surgery, 4) surgeons have not to rely on burning smell or colour change i.e. temperature related activities will be monitored by system 5) patient early rehabilitation.

5. Discussion

The experimentation performed and optimization indicates drill diameter (rank 1, contribution 90.71%) affects temperature rise most followed by feed rate (rank 2, contribution 5.69%) and spindle speed (rank 3, contribution 2.22). Anova analysis conducted with confidence level for 95% and significance level (α) =0.05.P value $< \alpha$, specifies the rejection of null hypothesis and acceptance of alternative hypothesis. The F-values for respective control factors represents the significance of that term.

Bone drilling is combination of drilling parameters [23] and specifications interaction in biological sphere. For successful conduction, this machining operation has to be conducted within a small temperature range. Though use of coolant and chips removal is recommended for temperature control, the reach of coolant at interface of tool and bone wall is not assured as the factors like stickiness of accumulated debris in drill bit helical grooves and its layered thickness over tool surface, the length of debris (similar to chips), rate of debris breakage are affected by bone mineral density (BMD), age and gender of patient and drill speed.at the same time a special attention has to be provided for micro cracks, which may cause fatigue damages and stress fractures.

6. Conclusion

Primarily drill diameter affects the heat accumulation whereas feed rate is the secondary factor and drill spindle speed is the last one. For all combinations of factors, with drill diameter of 2.5 and 3.2 mm temperature is found below 43^0c whereas for every combination with drill diameter 4.5 mm temperature is above 47^0c.Variation of drill speed as compared to drill diameter and feed rate is an effective way to control temperature. Another aspect is that apart from manual feed rate and drill tool diameter, drilling speed can be monitored and controlled with modifications in drilling machine. The proposed drilling machine with speed regulation will be helpful to orthopaedic surgeons by reducing dependency on conventional methods and also to patients by early healing and rehabilitation.

Ethical Approval: The sheep rib bone sample is collected from a local butcher after slaughtered during and for their routine trade practice and the sheep is not slaughtered intentionally for this research work. The authors are not related directly with any live animal hence ethical approval is not required.

References

[1] Matthews L S and Green C A 1984 The thermal effects of skeletal fixation pin insertion in bone *J. Bone Joint Surg.,* **66** pp 1077–83
[2] Eriksson R A and Albrektsson T 1984 The effect of heat on bone regeneration: an experimental study in the rabbit using the bone growth chamber *J. Oral Maxillofac. Surg.,* **42** pp 705-711
[3] Bonfleld W and Li C H 1968 The temperature dependence of the deformation of bone *J. Biomechanics,* **1,** 323-329 Pergamon Press Great Britain
[4] Lundskog J 1972 Heat and bone tissue *Scand. J. Plastic Recons. Surg. Sup*, **9**
[5] Tony M Keaveny 2004 Bone mechanics *standard handbook of Biomed. eng. design*
[6] Natali C and P Ingle et al 1996 Orthopaedic bone drills-can they be improved? Temperature changes near the drilling face *J. Bone Joint Surg.,* **78,** 357-362
[7] Augustin G and Davila S et al 2009 Determination of spatial distribution of increase in bone temperature during drilling by infrared thermography: preliminary report *Arch. Orthop. Trauma Surg.,* **129,** 703–709
[8] J Caeiro Potes et al 2008 The Sheep as an Animal Model in Orthopaedic Research *Experi. Patho.health sci.,* **2,** 29-32
[9] Lucia Martini et al 2001 Sheep Model in Orthopaedic Research: A Literature Review *Amer. Asso. Lab. Animal Sci.* August **51,** 292-299
[10] Jill E Shea Experimental Confirmation of the Sheep Model for Studying the Role of Calcified Fibrocartilage in Hip Fractures and Tendon Attachments *wiley-liss inc The anatom.record* , **266** 177–183
[11] Davidson SRH and James D F 2000 Measurement of thermal conductivity of bovine cortical bone *Med. Eng. Phys.,* **22,** 741–747
[12] Sharawy M and Misch C E et al 2002 Heat generation during implant drilling: the significance of motor speed *J. Oral and Maxillofac. Surg.,* **60,** 1160–1169
[13] Saha S and Pal S et al 1982 surgical drilling: design and performance of an improved drill *Trans.ASME J.Biomech. Eng.,* **104,** 245–52
[14] Augustin G and S Davila et al 2008 Thermal osteonecrosis and bone drilling parameters revisited *Arch. Orthop. & Trauma Surg.,* **128,** 71-77
[15] R V Dahibhate and S B Jaju 2020 Drilling in bone: Limitations and damage control by drilling specifications and parameters *j.mecha. of continua and mathe. sci.,* **15,** 153-161 doi.org/10.26782/jmcms.2020.04.00012
[16] Hillery M T and Shuaib I 1999 Temperature effects in the drilling of human and bovine bone *J. Mater. Process.Technol.,* **92,** 302–308
[17] Karaca F et al 2011 Influence of orthopaedic drilling parameters on temperature and histopathology of bovine tibia: An in vitro study *Med. Eng. Phys.,* **33,** 1221– 1227
[18] R V Dahibhate and S B Jaju 2019 Bone drilling parameters and necrosis: An in vitro study *springer nature Singapore pte Ltd* pp 599-606 doi.org/10.1007/978-981-13-6148-7_57
[19] Brisman D L 1996 The effect of speed, pressure, and time on bone temperature during the

drilling of implant sites *Int. J. Oral Maxillofac. Implants,* **11,** 35–37
[20] Roy R 2001 Design of experiments using the Taguchi approach: 16 steps to product and process improvement *John Wiley & Sons New York* ISBN 0471361011
[21] Toews A R and J V Bailey et al 1999 Effect of feed rate and drill speed on temperatures in equine cortical bone *Amer. J. Veter. Res.,* **60,** 942-944
[22] Nam O and Yu W et al 2006 Monitoring of bone temperature during osseous preparation for orthodontic micro-screw implants: effect of motor speed and pressure *Key Eng. Mater.* **(321–323),** 1044–47
[23] Ohashi H and Therin M et al 1994The effect of drilling parameters on bone *J. Mater. Sci: Mater. Med.,* **5,** 225–31

Effect of Nano SiO_2 as an Additive in Cotton Seed Oil

Dipak. S. Bajaj[1*], Dr.V.D. Wakchaure[2], Dr.R.S.Jahagirdar[3], Manali D. Bajaj[4]

[1]Reserach Scholar, [2]Professor, [3]Principal, [4]UG Scholar

[1,2]Amrutvahini College of Engineering, Sangamner, [MS], India-422 608.

[3] KG Reddy COE, Hyderabad, Telangana

[4] SVNIT, Surat

*E-Mail: dipakbajaj@yahoo.com

Abstract: Vegetable oil can be a deserving candidate as an eco-friendly lubricant because of its properties like biodegradability, non-toxicity, excellent lubricity, Viscosity Index and low volatility. But they have some unsatisfactory tribological properties. The lubricant used in this investigation is cottonseed oil, focused on the issues related to environmental pollution, toxicity and its cost. Even though cottonseed oil lubricant is one of the best choices for lubrication purpose, however the properties are slightly lesser than mineral oil. To improve the lubricant properties, nano Silicon dioxide additive is added in different concentrations. Four ball Tribotester is used for conducting anti-wear test according to ASTM standard D 4172. This paper describes the effect of Silicon dioxide nano additives on tribological properties of refined cottonseed oil. The refined cottonseed oil with 2 % of Silicon dioxide as additive shows good results which are very close to 20W50 oil.

Keywords: Vegetable oil, Wear Scar Diameter, Cottonseed oil, Silicon dioxide, 20W50, ASTM D 4172.

1. Introduction

Now days market demands for lubricants are increased environmental compatibility, reduced emissions with greater safety, and superior performance. Proper lubrication between the moving parts is very much essential for the effective and efficient working of an automobile at various operating conditions. Because of proper lubricant, the parts can slide smoothly over mating surfaces. This decreases loss of energy and also reduces wear and friction which has very importance in engines and drives. From last many years mineral oils have been used in engine as a lubricant. Now, the depletion of reserves of mineral oil and giving importance to the environment against pollution have changed the interest of researchers towards making of eco-friendly lubricants as a substitutes for mineral oils in industrial applications and also in engines [6]. Replacing petroleum base oils with biodegradable oils is one of the ways to reduce the adverse effects on the ecosystem caused by the use of lubricants. Vegetable oil based lubricants are having the most important properties such as high lubricity, high load carrying capacity, high flash point, high viscosity index, low toxicity, low emission into the environment as compared to synthetic and mineral oils [1,2,3]. However, the tribological properties of vegetable oil are poor and can be improved by adding suitable additives or by performing some chemical reactions [7, 8]. India has a great potential of producing edible and non-edible vegetable oil seeds. These oil seeds can be used as a most prospective source for producing vegetable oil based lubricants [3]. This experimentation will try to check the possibility of the replacement of mineral oil based lubricants with cottonseed oil based lubricants.

2. Vegetable Oils as Alternative Lubricants

Tribology is the branch of science which deals with study of interacting surfaces which are having relative motion between them. This brings all major disciplines together involving mainly wear, friction and lubrication between the matting surfaces. So the study of different parameters like friction, wear, and lubrication related to tribology became an important aspect of the research. The lubrication process of reducing friction and wear between moving or sliding surfaces, by the inserting lubricants in between them is called lubrication [2]. The purposes of lubrication are to reduce friction, wear and heat between contacting surfaces. It is also required to reduce oxidation, prevention from rust and reduce energy consumption [1, 2].

2.1 Advantages and Disadvantages of Vegetable Oils

Vegetable oils have several advantages and also some limitations when considered for automobile, industrial and machinery lubrication because of their good inherent qualities. Table 1 summarizes the some of the advantages and disadvantages of vegetable oils as compared to mineral oils [1, 2, 3].

Table 1: Advantages and Disadvantages of vegetable oils

Advantages	**Disadvantages**
Fast biodegradation	Poor Oxidation Stability
Less toxic	Limitations in Operating Temperature
Renewable and clean	Unpleasant smell
High VI	High pour point
Low volatility	
Higher lubricity	
Higher flash points	
Good boundary lubrication	
Reduced engine emissions	
Disposal at minimal expense	

2.2 Properties of Vegetable Oils

Vegetable based lubricating oils have many useful physio-chemical properties which are listed in Table 2 [5].

Table 2: Properties of vegetable oils

Properties → Vegetable oils↓	**Kinematic Viscosity (at 40°c) cSt**	**Pour point (°c)**	**Cloud point (°c)**	**Flash point (°c)**	**Density (kg/l)**
Corn oil	34.9	-40.0	-1.1	277	0.9095
Linseed oil	22.2	-15.0	1.7	241	0.9236
Peanut oil	39.6	-6.7	12.8	271	0.9026
Rapeseed oil	37.0	-31.7	-3.9	246	0.9115
Soya bean oil	32.6	-12.2	-3.9	254	0.9138
Sunflower oil	33.9	-15.0	7.2	274	0.9161
Palm oil	39.6	-	31.0	267	0.9180
Cotton seed oil	**33.5**	**-15.0**	**1.7**	**234**	**0.9148**

3. Experimentation

3.1 Selection of Oil and nano additive

Cottonseed oil, in India, is available on large scale, mainly in Gujarat and Maharashtra state. Moreover, the cost of cottonseed oil is very less as compared to others. Silicon dioxide is used as a nano additive and added with different concentrations in the refined cotton seed oil. The specifications obtained from supplier for Silicon dioxide are Purity: 99.5% APS: 15-20 nm and Density: 2.4 g/cm^3. The present work is an attempt to touch the potential of refined cottonseed oil as an alternative lubricant compared to mineral oils. Generally, SAE multi-grade oil i.e. SAE 20W50 is used as a lubricant in multi-cylinder engine to have proper lubrication. Table 3 shows the properties of SAE 20W50 and refined cotton seed oil. These properties of RCSO are tested at CHEMTECH Laboratories, Pune and of SAE 20W50 oil is taken from manufactures catalogue.

Table 3: Properties of lubricating oil

Lubricant→ **Properties ↓**	**SAE 20W50 oil**	**RCSO**	**Test Method**
Specific gravity	0.8954	0.9040	ASTM D 4052 2016
Kinematic Viscosity (40°c) cSt	174.8	32.78	ASTM D 7042 2016
Kinematic Viscosity (100°c) cSt	19.1	15.80	ASTM D 7042 2016
Viscosity index	124	181.67	ASTM D 2270 2016
Flash point (°C)	229	290-292	ASTM D 92 2016 b
Fire Point (°C)	--	304-306	ASTM D 92 2016 b
Pour point (°C)	-30	-20	ASTM D 97 2017 b

3.2 Sample Preparation

Silicon Dioxide as anti-wear additive is added at 1%, 1.5% and 2% wt. concentrations in refined cotton seed oil (RCSO) to prepare the test samples. The sample is prepared of 30 ml by volume. The weight of the nano additive is measured on a digital weighing machine. The magnetic stirrer with hot plate is used for dispersion of additive in the oil. The dispersion process with the help of magnetic stirrer improves the dispersion stability. The sample calculation is described below. All other sample calculations are also done in the same way.

3.3 Sample calculation for RCSO + 1% SiO_2

Volume of Sample	= 30 ml
Volume of RCSO	= 30 ml
Weight of RCSO	= Volume × Density
	= 30 × 0.9060 gm/ml
	= 27.18 gm
Weight of SiO_2	= 27.18 × 1/100
	= 0.2718gm
Volume of SiO_2	= weight / density
	=0.2718 gm /2.4 (gm/cm^3)
	= 0.1133 ml
Actual volume of RCSO	= 30 – 0.1133
	= 29.8867 ml

3.4 Tribological Testing on Four Ball Tribotester

The wear preventive characteristics of lubricant were tested as per ASTM D 4172 standard. The four-ball tribotester TR-30L-IAS, manufactured by DUCOM, Bangalore, India is used to evaluate wear preventive characteristics of oil. The trobotester uses four balls where three balls are kept at the bottom and one on top. The bottom three balls are fastened firmly in a ball pot which is containing the lubricating oil under test. The top ball is rotated at the desired speed, while the bottom three balls are pressed against it. The lubricant under test is characterized by measuring the wear scar formed on the surface of balls. This wear scar is measured using an image acquisition system [10]. The Four-ball tribotester is shown in Fig. 1 and the testing conditions as per ASTM D 4172 are as shown in Table 4.

Table 4: Wear Test Conditions

Parameters	Condition
Load	392 ± 2 N
Temperature	75 ± 2 °C
Speed	1200 ± 60 rpm
Time	60 ± 1 min

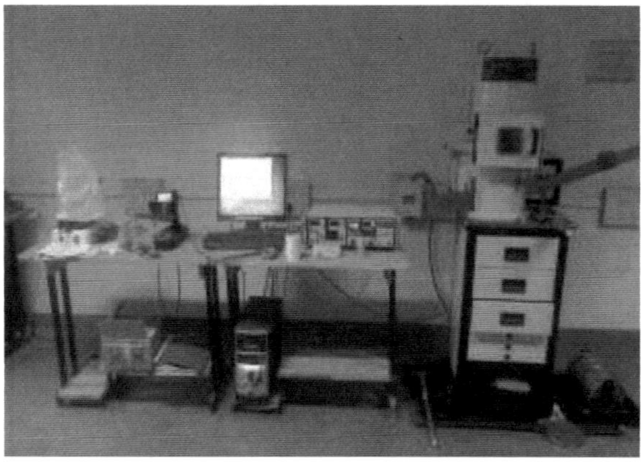

Figure 1: Ducom TR 30 L Four Ball Test Rig.

Testing balls are made of chrome alloy steel (AISI standard E-52100), with a diameter of 12.7 mm and a roughness value of Ra = 0.035 mm. Acetone is used as a cleaning fluid, which is nontoxic and avoiding transfer of test oil from one test to the next. It is also not contributing in wear or anti-wear of the test lubricant.

4. Results and Discussion

The results obtained from Four-Ball Tribotester for wear tests of testing samples with and without additives are discussed further.

4.1 Friction and Wear Behavior

The wear characteristics (Wear Scar Diameter) of lubricating oil were measured using the image acquisition system. The average values of the wear scar diameters and coefficients of friction for all samples of SAE 20W50 oil and refined cottonseed oil are as shown in Table 5.

Table 5: Average values of coefficient of friction and wear scar diameter for test oils.

Sample Oil	Average WSD (μm)	Coefficient of Friction
SAE 20W50 oil	530	0.08839
Cottonseed oil (CSO)	866	0.07720
Refined cottonseed oil (RCSO)	696	0.09259
RCSO+1% Silicon Dioxide	624.33	0.09822
RCSO+1.5% Silicon Dioxide	565.66	0.08870
RCSO+2% Silicon Dioxide	558	0.08947

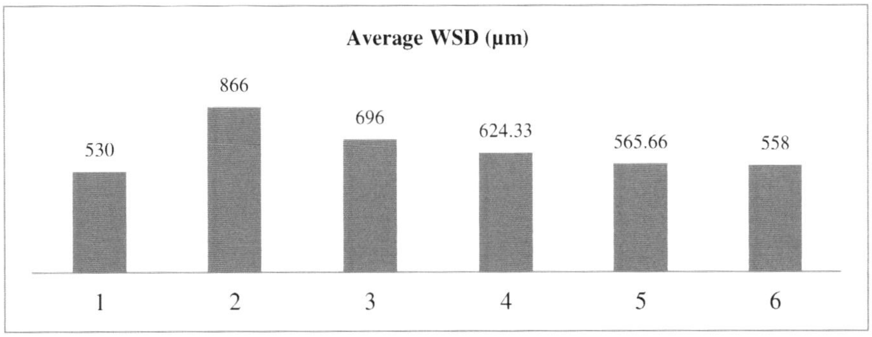

(a)

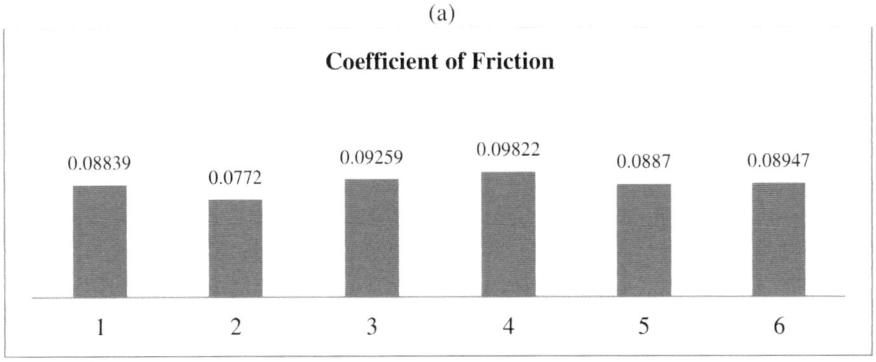

(b)

Figure 2: (a) WSD and (b) COF for different samples

B1 B2 B3

Figure 3: Scar images for SAE 20W50 Oil

B1 B2 B3

Figure 4: Scar images for Refined CSO

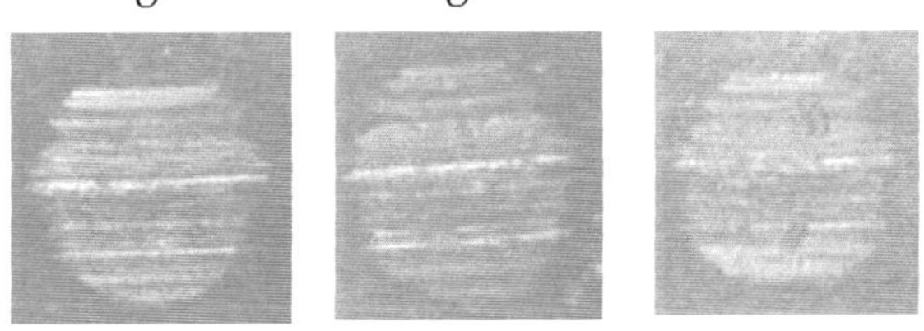

B1 B2 B3

Figure 5: Scar images for RCSO + 2% wt. SiO_2

IOP Conf. Series: Materials Science and Engineering 1004 (2020) 012016 doi:10.1088/1757-899X/1004/1/012016

From the results shown in the table 5, Figure 2, 3, 4 and 5, it is observed that coefficient of friction for pure cottonseed oil is less than that of the coefficient of friction of SAE 20W50 oil, but the scar diameter is higher. But wear scar diameter and coefficient of friction of refined cottonseed oil is larger than SAE 20W50 oil. This increase in wear scar diameter is because of the continuous removal of metallic soap film. This film is formed because of the reaction of the oil with the metallic surface during relative motion. This metallic film is continuously reforming by further chemical reactions. The shear strength of the metallic soaps is low, so as a result, the coefficient of friction for vegetable oil is also low. From the Table 5, further it is observed that the wear scar diameters for 1%, 1.5%, and 2% silicon dioxide additive wear scar diameter is lesser than refined cottonseed oil. The wear scar diameter of RCSO is decreasing with increasing the concentration of SiO_2 additive. The 2% SiO_2 with refined cotton seed oil shows the lowest value of wear scar diameter. This is because SiO_2 forms a reactive film layer that prevents direct contact between metal surfaces. The protecting layer may act as a cushion between the metal and it prevents the formation of corrosion wear on metal surfaces. The coefficient of friction remains almost same for RCSO + 2% SiO_2 as like 20W50 oil. Figure 6 shows the graph between Frictional Torque and time. It shows that it remains almost constant when SiO_2 added in refined cotton seed oil.

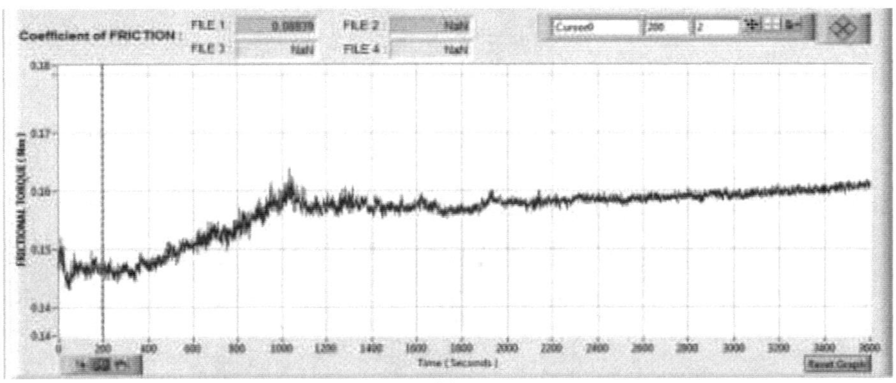

(a)

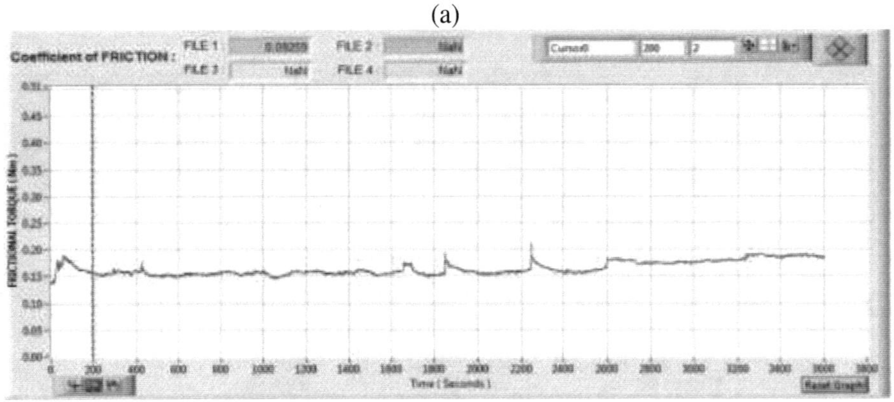

(b)

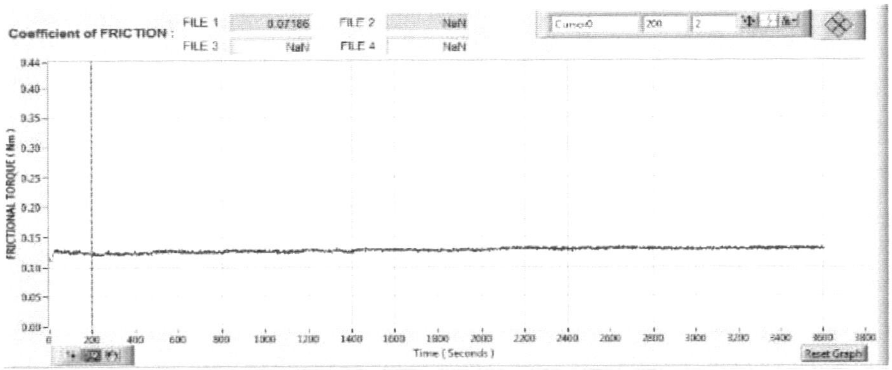

(c)

Figure 6: Graph of frictional torque (Nm) Vs time (sec) for (a) SAE 20W50 oil (b) RCSO (c) RCSO + 2% SiO_2

5. Conclusions

India is one of the largest countries in production of oilseeds in the world with its rich agro-ecological diversity. So, India can grab this opportunity to expand the market share of lubricants based on vegetable oil. Vegetable oil-based lubricants are biodegradable and biodegradability is one of the major required properties of lubricant in case of automobile applications. From anti-wear test carried out on Four-Ball Tribotester, the wear scar diameter of refined cottonseed oil is more; it can be improved by adding anti-wear additives like Silicon Dioxide. The WSD of RCSO+2% SiO_2 is lower and very close to 20W50. At the same time there is no much difference in the COF. It can be concluded that the cottonseed oil can become an alternative for mineral oil by adding some additives.

Acknowledgement

The author thanks BCUD, Savitribai Phule Pune University, Pune for providing financial assistance under Research Promotion Scheme.

References

[1] H M Mobarak, E Niza Mohamad,H H Masjuki, M A Kalam, K A H Al Mahmud, M Abibullah, A M Ashraful 2014 The prospects of biolubricants as alternatives in automotive applications *Journal of Renewable and Sustainable Energy Reviews*, **33**, 34-43.

[2] Jan C J Bart, Emanuele Gucciardi, Stefano Cavallaro 2013 Biolubricants Science and Technology *Woodhead Publishing Limited,* **46,** 52-67.

[3] Anjana Srivastava, Preeti Sahai 2013 Vegetable oils as lube basestocks: A review *African Journal of Biotechnology* **12(9)**, 880-89.

[4] Salimon J, Abdullah B, Yusop R M, Salih N 2014 Synthesis, reactivity and application studies for different biolubricants *Chemistry Central Journal*, **8(1),** 1-11.

[5] Avinash Kumar Agarwal 2007 Biofuels (alcohols and biodiesel) applications as fuels for internal combustion engines *Progress in Energy and Combustion Science*, **33(3)**, 233–271.

[6] Sachin M Agrawal, Subhash Lahane, N. G. Patil, P. K. Brahmankar 2014 Experimental investigations into wear characteristics of M2 steel using cotton seed oil *Procedia Engineering*, **97**, 4-14.

[7] K Balamurugan, N Kanagasabapathy, K Mayilsamy 2010 Studies on soyabean oil based lubricant for diesel engines *Journal of Scientific and Industrial Research*, **69**, 794-797.

[8] N H Jayadas, K Prabhakaran Nair, Ajithkumar G 2007 Tribological evaluation of coconut oil as an environment-friendly lubricant *Tribology International*, **40(2)**, 350-354.

[9] Chacko Preno Koshy, Perikinalil Krishnan Rajendrakumar, Manu Varghese Thottackkad 2015 Evaluation of the tribological and thermo-physical properties of coconut oil added with MoS_2 nanoparticles at elevated temperatures *Wear*, pp. **330-331**, 288-308.

[10] Ducom Four Ball Tester Product Information Nov 2016 (https://ducom.com/test-instruments/lubricant-testers/four-ball-tester)

IOP Conf. Series: Materials Science and Engineering 1004 (2020) 012017 doi:10.1088/1757-899X/1004/1/012017

Unidirectional dry sliding wear analysis of ZE41 magnesium alloy

B Kaushik[1*], S Anand Kumar[1], T R Prabhu[2] and A Moganraj[3]**
[1]Department of Mechanical Engineering, Indian Institute of Technology Jammu, J & K, India
[2]Scientist, Defence Research and Development Organisation, Bangalore, India
[3]School of Mechanical Sciences Vellore Institute of Technology Vellore, Tamil Nadu, India
E-mail: *kaushik.brahmansh@gmail.com and **anandkumarait@gmail.com

Abstract. Magnesium alloys have many characteristics like good castability, high strength to-weight ratio and good amenability. The ZE41 magnesium alloy consists of zirconium (Zr), zinc (Zn) and rare earth elements. Magnesium alloys have many application in aerospace, automotive industries and biomedical implants. Microstructure is shown for ZE41 magnesium alloy Wear rates are measured for ZE41 magnesium alloy using pin-on-disk experiment setup. Rate of wear and coefficient of friction is measured for 10N-20N range at sliding velocity of 0.5 m/s with constant sliding distance of 300 meters. It is observed that load is not affecting coefficient of friction significantly but there is decrement in the value of CoF when load is increased from 10N to 15N. Wear behaviour in initial condition is different but it is converging to general which means more the load more will be the wear. Wear mechanism is abrasion for low loads then slightly changes to delamination.

1. Introduction

Magnesium alloys have many applications in industries where reduction of weight is the main concern such as automobile industries, aerospace industries and bio medical implants. Magnesium based alloys are used as degradable biomaterial implants in humans for temporary implants applications [1]. Magnesium alloys are suitable to these light weight applications due to its high strength-to-weight ratio. These alloys also have high corrosion resistance, low density, dimensional stability and castability. Casted product of ZE41 can be altered in grain size with controlled cooling method [2]. Magnesium ZE41 alloy is commonly used for gravity casting for limited creep resistance corrosion properties due to the rare earth elements in it. Composition of ZE41 magnesium alloy is mostly consist of Magnesium (Mg) while other parts are Zinc (Zn) as 3.12%, Zirconium (Zr) as 0.54%. Rare earth elements as Gadolinium (Gd), Neodymium (Nd) etc. are 0.75-1.75% (wt) [3].The rare earth elements helps in alloy melting, refining in microstructure and increase anti-oxidizing properties whereas Zinc(Zn) improves ductility and also strength at high temperature [4]. The tribological tests are carried out to understand the insight of wear behaviour of ZE41 magnesium alloy for constant sliding velocity and different load conditions. It should be noted that improving the surface properties are found to be effective way to wear resistance of the engineering materials particularly surface hardness. In this regard, several surface modification techniques were employed to improve the tribological performance of non-ferrous and ferrous alloys [5-12]. As in most of the application of magnesium alloys like in aircraft and automobile industries analysis of sliding wear behaviour is very important because of load and constant contact between two components. In the present work, effect of normal load on the friction curves and dynamic wear loss under unidirectional sliding wear mode is investigated.

2. Experimental methods

Wear experiments are performed on pin-on-disk tribology tester with ZE41 magnesium alloy as pin in 3mm X 3mm X 6mm size. The counter-body was EN-31 hardened to 58-62 HRc steel disk with 165mm diameter. For experiments different load condition of 10N, 15N and 20N. All the samples are rum on test for 600 Sec with constant sliding velocity of 0.5m/s which results in sliding distance of 300 meters. Design of experiment See Table (1). All specimens and tribometer disk are cleaned with acetone for elimination of any undesired particles or debris.

Table 1: Experiment Design Table

Sample No.	Load (N)	Time (sec)	Sliding Velocity (m/s)
Sample No. 1	10	600	0.5
Sample No. 2	15	600	0.5
Sample No. 3	20	600	0.5

3. Results and Discussion

3.1. Microstructural Analysis

For microscopic analysis sample surface is grounded with 100 grit to 2000 grit emery paper. After that polished with diamond paste. To develop microstructure picric acid, acetic acid, ethanol, distilled water is used as an etchant for ZE41 Magnesium alloy. Observations are carried under optical microscope at optical magnification. Fig. 1 shows the microstructure was almost having an equiaxed grains with α-Mg with eutectic precipitates boundaries of grains. Similar microstructure was observed in study [13].

(a) At 139x total magnification (b) At 693x total magnification

Figure 1: Microstructure of ZE41 magnesium alloy

3.2. Hardness Analysis

Hardness measurement was carried out using micro hardness tester for 500g load and dwell time of 10 seconds using diamond indenter. A total of 5 hardness measurements are performed at different locations on the sample. The averaged value of hardness for ZE41 magnesium alloy is 68.788 ± 6.032 $HV/_{0.5}$. Rare earth elements such as Neodymium(Nd) has highest solid solubility in magnesium and helps in the improvement of strength with increase in (Nd) as it contributes to both solid solution hardening capabilities as well as for precipitation hardening whereas Gadolinium(Gd) helps in elongation to fracture compared to other metallic alloys [14].

IOP Conf. Series: Materials Science and Engineering 1004 (2020) 012017 doi:10.1088/1757-899X/1004/1/012017

3.3. *Coefficient of Friction Analysis*

Coefficient of friction is calculated for each data point of friction force value using formula (CoF = Friction Force/Load Applied) and plotted against time for each load condition as mentioned in Table 1. All the plots are shown in the Fig. 2.

The effect of normal load on coefficient of friction shown figure 2d. It shows similar values of coefficient of friction for all load conditions. In a study on magnesium alloy AZ71E investigated by Huang Weijiu [15] shows that the coefficient of friction decreases as the load is increased. This same is observed in the present study for ZE41 magnesium alloy in which coefficient of friction is decreased on the increment of load from 10N to 15N.

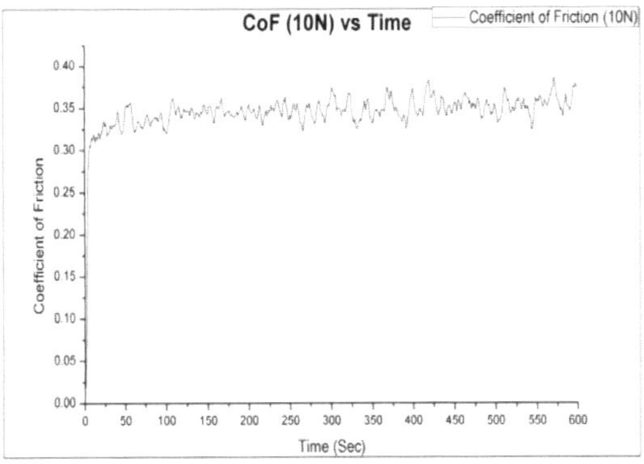

(a) Graph of 10N Load for CoF vs Time

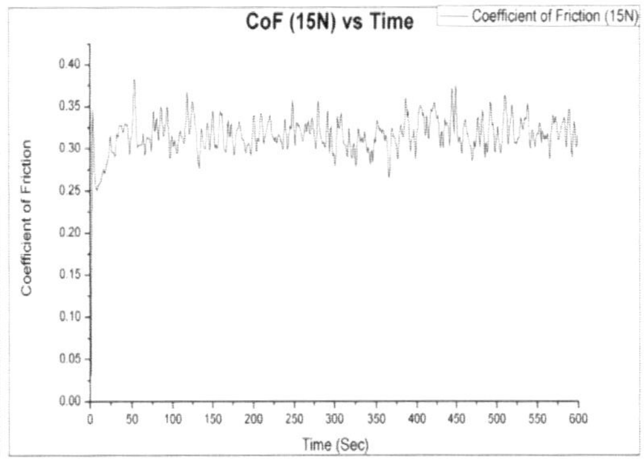

(b) Graph of 15N Load for CoF vs Time

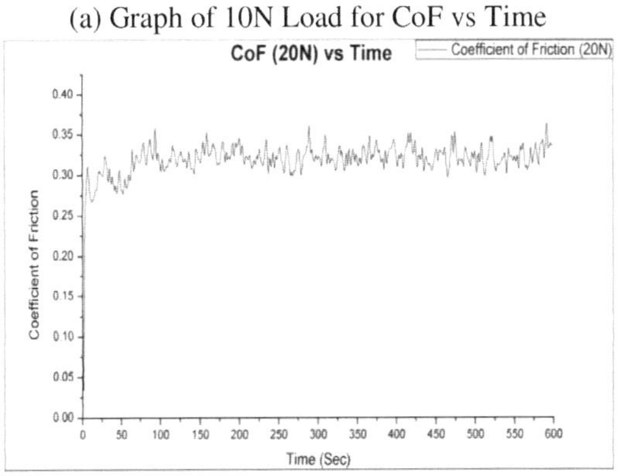

(c) Graph of 20N Load for CoF vs Time

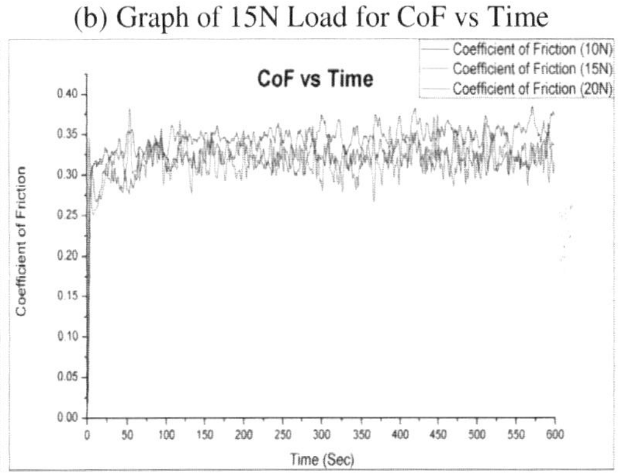

(d) Combined graph of Load for CoF vs Time

Figure 2: Graphs of Coefficient of Friction vs Time for various load conditions

Table 2: Mean Coefficient of Friction for different normal Load

Sample No.	Load (N)	Time (sec)	Sliding Velocity (m/s)	CoF (Mean)
Sample No. 1	10	600	0.5	0.345
Sample No. 2	15	600	0.5	0.316
Sample No. 3	20	600	0.5	0.319

3.4. *Wear Analysis*

Wear behaviour for different load is shown in Fig. 3. In initial stage sliding wear rate is highest for 10N load compared with 15N and 20N (See Table 2). As sliding distance increase which means time increase it shows similar slopes as of 15N and 20N. It is interesting to note that in beginning of the experiment

20N load sample shows least wear rate but it crosses 15N sample at 257 sec roughly also crosses 10N sample at the end of the experiment. Huang weijiu [15] found that sliding mechanism changes on the application of load, with increase in load the mechanism changes to delamination and melting from abrasion. Wear rate comparison is performed by using slope analysis (See Table 3). It shows maximum wear rate in the beginning is for 10N load but as sliding distance increases, wear rate is maximum for 20N load compared to 10N load. This behaviour is also observed by A.J Lopez [16] for sliding velocity 0.5m/s wear rate shown as mass loss/sliding (g/m) increases with increase in load. Similar wear rate trend is also observed in study performed by Chiranth [13] that on increase of load wear rate increased.

Table 3: Data of Wear with slope change in linearly fitted data

	slope		
Time (Sec)	10N	15N	20N
0-15	5.3823	1.5981	1.0205
100-600	0.2906	0.3736	0.4481

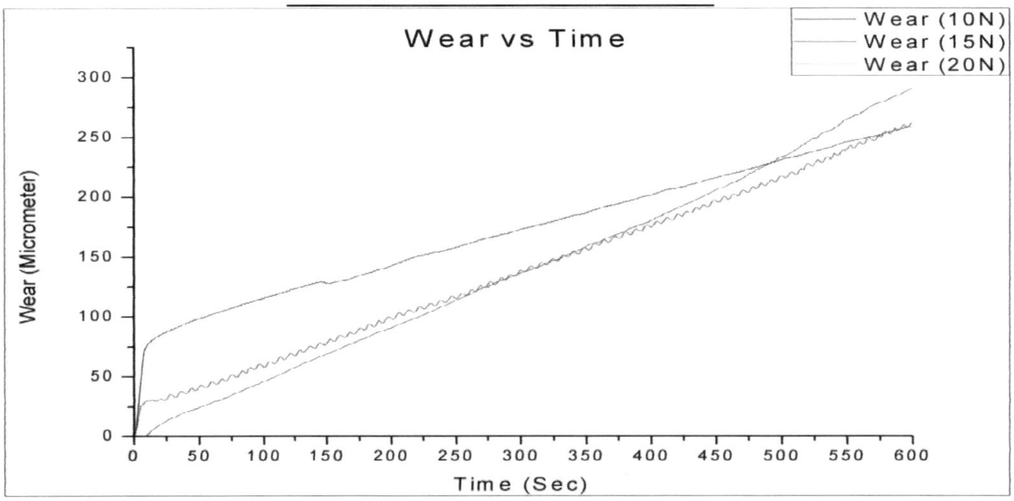

Figure 3: Graph of Sliding Wear vs Time of Loads

4. Conclusion

In the present study, the material ZE41 magnesium alloy investigated has equiaxed grains and grain boundaries are precipitates of rare earth elements. It observed that at constant sliding velocity condition for different loads, the coefficients of friction shows decrease in the value on increment of load suddenly for 10N to 15N but for 15N and 20N almost similar value is observed. Sliding wear in the initial stage is different, but in the steady state, it observed a progressive trend of wear, which is more the load more the wear rate. It is presumed that at sliding velocity used in the present study mainly abrasion wear mechanism could have played as key role however wear mechanism slightly changes from abrasion toward delamination and melting when load is increased.

References

[1] Witte F, Hort N, Vogt C, Cohen S, Kainer K U, Willumeit R and Feyerabend F 2008 *Current Opinion in Solid State and Materials Science,* **12,** 63–72

[2] Kaushik B, Rao S, Kumar S and Kushwaha A 2019 *Manufacturing Technology Today,* **18,** 51–56

[3] Ch S, Anuradha N, Geeta Krishna P and Reddy B 2017 *International Journal of Mechanical Engineering and Technology,* **8,** 722–728
[4] Cai S, Lei T, Li N and Feng F 2012 *Materials Science and Engineering: C,* **32,** 2570–2577
[5] Anand Kumar S, Ganesh Sundara Raman S, Sankara Narayanan T and Gnanamoorthy R 2012 *Surface and Coatings Technology,* **206,** 4425 – 4432 ISSN 0257-8972
[6] Anand Kumar S, Ganesh Sundara Raman S, Sankara Narayanan T and Gnanamoorthy R 2013 *Tribology International,* **57,** 107 – 114 ISSN 0301-679X
[7] Kumar S, Pradhan S, Ganesh Sundara Raman S and Gnanamoorthy R 2013 *Proceedings of the Institution of Mechanical Engineers, Part J: Journal of Engineering Tribology,* **228,** 454–462
[8] Kumar S A, Sundar R, Raman S G S, Kumar H, Gnanamoorthy R, Kaul R, Ranganathan K, Oak S M and Kukreja L M 2012 *Tribology Transactions,* **55,** 615–623
[9] Kumar S A, Sundar R, Raman S G S, Gnanamoorthy R, Kaul R, Ranganathan K and Bindra K S 2017 *Proceedings of the Institution of Mechanical Engineers, Part J: Journal of Engineering Tribology,* **231,** 1276–1288
[10] Kumar S, Ganesh Sundara Raman S, TSN S N and Gnanamoorthy R 2012 *Advanced Materials Research,* **463-464,** 316 – 320
[11] Kondaiah V, Pavanteja P, Afzal Khan P, Anannd Kumar S, Dumpala R and Ratna Sunil B 2017 *Materials Today: Proceedings,* **4,** 6671 – 6677 ISSN 2214-7853 international Conference and Expo on Magnesium, iMagCon2016, VIT University, Chennai Campus, February 4 – 6, 2016
[12] Shen X, Shukla P, Subramaniyan A K, Zammit A, Swanson P, Lawrence J and Fitzpatrick M E 2020 *Optics Laser Technology,* **131,** 106446 ISSN 0030-3992
[13] BP C, Siddaraju C, Mishra R, Sasikumar R, Sathiskumar R and Prabhu T 2019 *Materials Science Forum,* **969,** 86–92
[14] Tekumalla S, Seetharaman S, Almajid A and Gupta M 2014 *Metals,* **5,** 1–39
[15] HUANG W j, LIN Q and LIU C l 2012 *Transactions of Nonferrous Metals Society of China* **22** 2057–2065
[16] L´opez A J, Rodrigo Herrero P, Torres B and Rams J 2011 *Wear,* **271,** 2836–2844

Effect of Impactor velocity and boundary condition on low velocity impact finite element modelling of CFRP composite laminates

K. R. Kaware[1*], Dr. M. S. Kotambkar[2]
[1]Ph.D research scholar, Visvesvaraya National Institute of Technology, Nagpur MH
[2]Assistant Professor, Visvesvaraya National Institute of Technology, Nagpur MH
[*]Email: kiran.kaware@gmail.com

Abstract: The demand for Composite structures is increasing in different engineering fields such as automobile, aircraft, civil because of its higher specific mechanical properties. It's very important to consider failure modes like fiber breakage, delamination, matrix cracking, etc. when it is impacted by a foreign object such as dropping of tool (impact type events). Susceptibility of damage due to low velocity impact is more in the laminated composite structures than metal. It is important to study the response of composite for varying velocities with different boundary conditions in case of low velocity impact analysis. Thus, this paper presents the combined effect of different boundary conditions with varying velocities of impactor on low velocity impact modelling of CFRP composite laminates. Numerical analysis is performed on T700GC/M21 material with layup $[0_2/45_2/90_2/-45_2]_s$ to study the force-time/displacement curve. Numerical analysis is performed using ABAQUS 6.14. First, Numerical analysis is carried out for impactor velocities varying from 2.5 m/s to 10 m/s. In ABAQUS/Explicit, two boundary conditions i.e. simply supported edges boundary condition SSSS and Clamped supported edges, CCCC are used to study the force-time/displacement curve. Out of that numerical results of impactor velocity 5 m/s with boundary condition SSSS is validated with results of literature by studying force-time/displacement curves. The numerical findings revealed that; as impactor velocity increases, the force and displacement also increases. Also, current simulation results are closely matching with the existing numerical and experimental data.
Keywords: Impactor velocity, low velocity impact, boundary condition, force-time/displacement, finite element.

1. Introduction:

The demand for Composite structures is increasing in different engineering fields such as automobiles and aircraft due to their inherently higher specific mechanical properties. The most important features of composites are higher strength, high stiffness, low density, and shear properties, etc. These composite laminates or structures are often subjected to dynamic impact loads during their use. Plastic deformations, matrix cracking, delaminations, Fiber breakage are some effects that should be taken care of when there is a low velocity impact (tool drop while maintenance, bird strike, etc.) on composite structures.

Low velocity impact and high velocity impact are the two main classifications of the impact problem of the composite while studying the damage mechanisms. To study damage mechanism, usually low velocity impact analysis is done due to its lightweight design. The damage due to low velocity impact is not clearly visible to naked eyes when the composite laminate is impacted by some foreign object.

For composite structures under the impact, the primary failure mechanism includes transverse matrix cracking, longitudinal fibre breakage, delamination also interactions.

Finite element method (FEM) is an essential tool available for the numerical analysis of composite structures. It is beneficial to study failure properties while designing the composite. Numeral analysis in ABAQUS includes a series of complicated numerical issues like geometry, material properties, boundary conditions, frictional contact, failure law, and time-stepping, which are essential while modelling low velocity impact.

Zuleyha Aslan et al. [2] studied the effect of low velocity impact loading on laminated composite. The in-plane dimensional effect of CFRP laminated composite evaluated for impact loading. Numerical simulation is done using 3D IMPACT for calculating contact forces & stresses during impact. Volnei Tita et al. [3] presented a study on thin disks of carbon fiber composite with epoxy resin for low velocity impact. An investigation is done to see the effect of energy impact and stacking sequence using force-time/displacement curves by the NDE technique.

D. Fang and E. Aymerich [4] presented a damage model that is based on continuum damage mechanics. The study carried out for the prediction of different failure mechanisms and structural response of composite laminates low velocity impact loading. Shi-Xu Wang et al. [11] investigated low velocity impact characteristics and residual tensile strength of CFRP laminates numerically (using ABAQUS/Explicit) and experimentally.

Jikui zhang and xiang zhang [5] presented an approach to predict the damage mode and impact force in laminated composite under low velocity impact. N. Hongkarnjanakul et al. [6], Y. Shi et al.[7], Wei Tan et al. [8] used ABAQUS for low velocity impact finite element modelling and analysis of CFRP composite laminates.

In the above literature, the impact response of the composite structure is found out by different EF software but the combination of velocity with boundary conditions is less evolved. So, This paper presents the combined effect of different boundary conditions with varying impactor velocities on low velocity impact modelling of CFRP composite laminates. The stacking sequence considered for analysis is $[0_2/45_2/90_2/-45_2]_s$ having 8-double ply, quasi-isotropic, and mirror-symmetric laminated plate of material T700GC/M21 (Carbon/Epoxy). The studies carried out on plate dimension 100mm×150mm with 4mm thickness on laminate. Impactor is of 16 mm diameter and having mass 2 Kg. First, Numerical analysis is carried out for impactor velocities varying from 2.5 to 10 m/s with the step of 2.5 for two boundary condition i.e., simply supported edges boundary condition SSSS and Clamped supported edges, CCCC to study the force-time/displacement curve. Out of that numerical results of impactor velocity 5 m/s with boundary condition, SSSS is validated with the results of P. F. Liu et al. [1] by studying force-time/displacement curves.

Properties for T700GC/M21 composite laminates

Layer Properties:-

$E_1 = 130$ GPa, $E_2 = E_3 = 7.7$ GPa, $G_{23} = 3.8$ GPa, $G_{12} = G_{13} = 4.8$ GPa, $\nu_{12} = \nu_{13} = 0.33$, $\nu_{23} = 0.35$, $X^T = 2080$ MPa, $X^C = 1250$MPa, $Y^T = 60$MPa, $Y^C = 110$ MPa, $Z^T = Z^C = 290$ MPa, $S_{12} = S_{23} = S_{13} = 110$ MPa, $\Gamma_{11}{}^T = 133$ N/mm, $\Gamma_{11}{}^C = 40$ N/mm, $\Gamma_{22}{}^T = 0.6$ N/mm, $\Gamma_{22}{}^C = 2.1$ N/mm

Interface properties:-

$G_1{}^C = 0.6$ N/mm, $G_2{}^C = G_3{}^C = 2.1$ N/mm, $\eta = 1.45$, $T_1{}^0 = T_2{}^0 = T_3{}^0 = 30$MPa

2. Finite element modeling of impact problems of composites

ABAQUS 6.14 is used for Low velocity impact Finite element modelling problem. Layup parameters, material properties, interactions between lamina to lamina, and laminate to impactor, impact velocity, the meshing of laminate and impactor, different boundary conditions are provided using ABAQUS. The reduced integration with hourglass control element SC8R used for the laminate. The 0.3 friction contact coefficient is considered. Material T700GC/M21 is used for layup $[0_2/45_2/90_2/-45_2]_s$ and material parameters are given in Table. The geometry size for laminate is 100mm×150mm×4mm and ply thickness is 0.25mm. The mass & diameter of the impactor are 2kg and 16mm respectively.

In the following, the effect of different impact velocities with boundary conditions are discussed. Also, validation of existing results is done.

Figure 1: FE impact analysis model

2.1 Effect of Impactor velocity for two boundary conditions

In this section, the impact analysis responses for velocities varying from 2.5 to 10m/s with two boundary conditions are presented.

2.1.1 Impactor Velocity 2.5 m/s

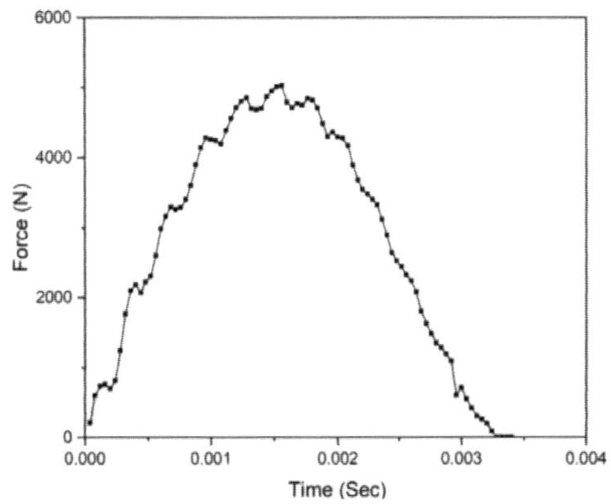

Figure 2: Impact force vs time for 2.5 m/s vel. with clamped supported edges, CCCC

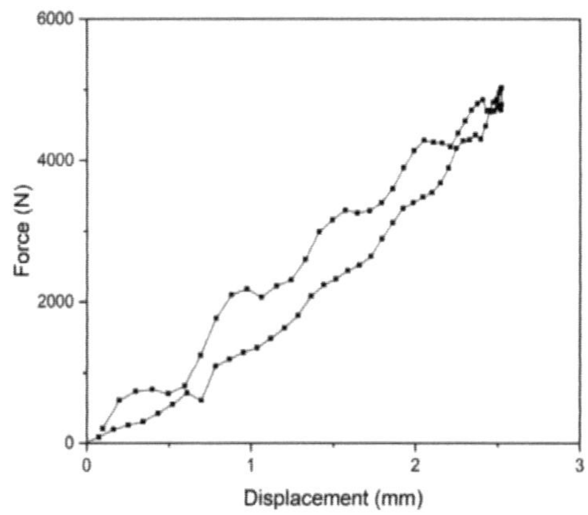

Figure 3: Impact force vs displacement for 2.5 m/s vel. with clamped supported edges, CCCC

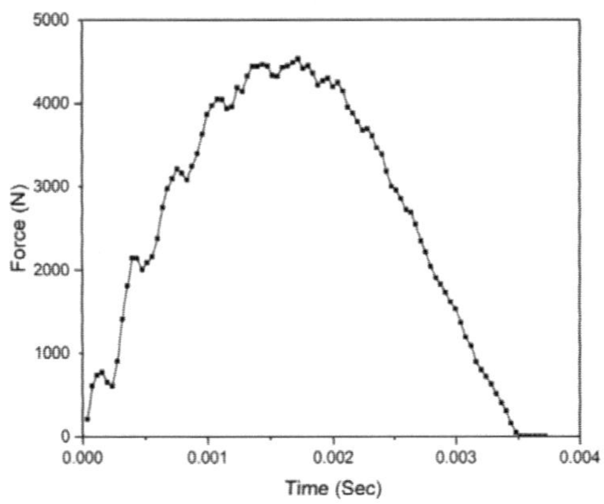

Figure 4: Impact force vs time for 2.5 m/s vel. with simply supported edges, SSSS

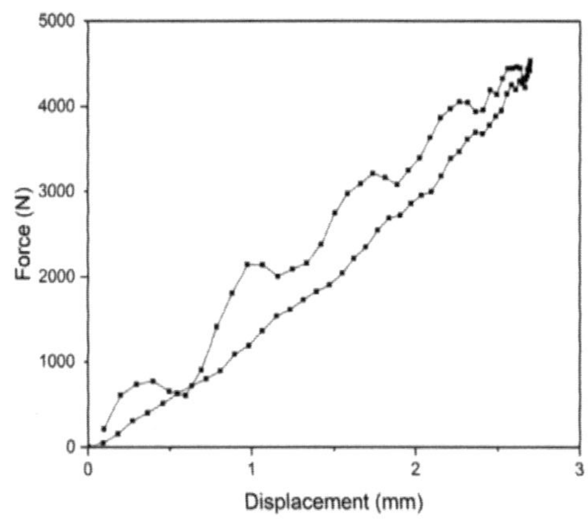

Figure 5: Impact force vs displacement for 2.5 m/s vel. with simply supported edges, SSSS

2.1.2 Impactor Velocity 5 m/s

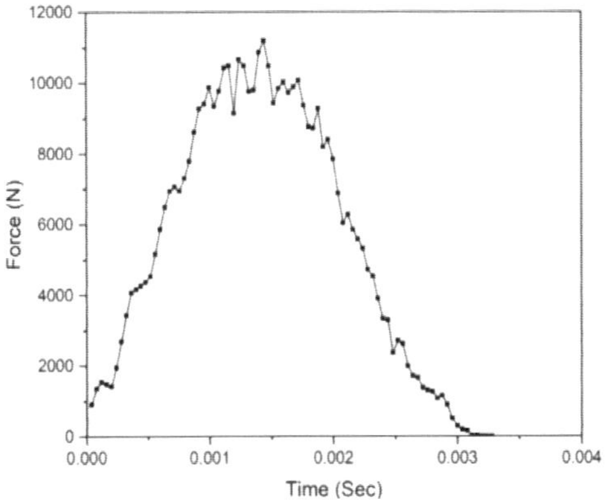

Figure 6: Impact force vs time for 5 m/s vel. with clamped supported edges, CCCC

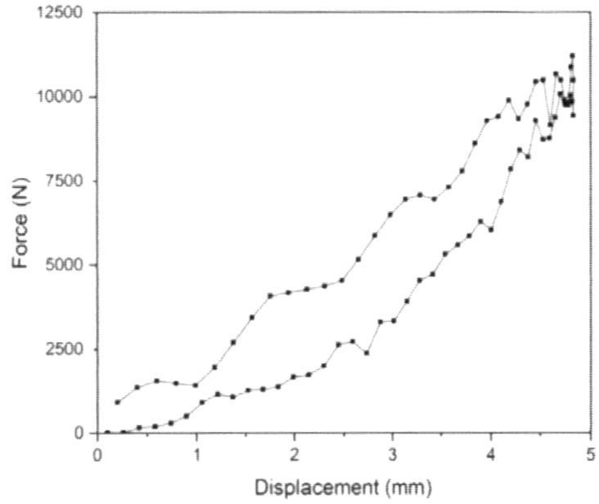

Figure 7: Impact force vs displacement for 5 m/s vel. with clamped supported edges, CCCC

2.1.3 Impactor Velocity 7.5 m/s

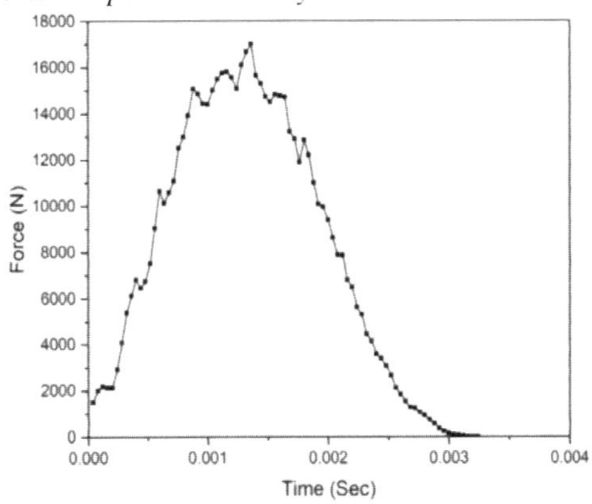

Figure 8: Impact force vs time curve for 7.5 m/s vel. with clamped supported edges, CCCC

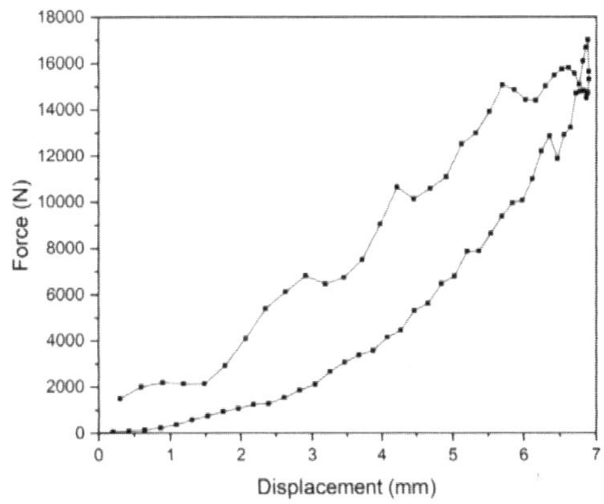

Figure 9:Impact force vs disp. curve for 7.5 m/s vel.with clamped supported edges, CCCC

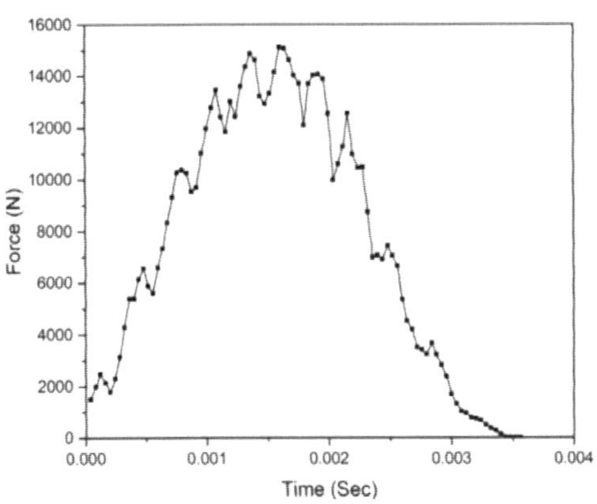

Figure10: Impact force vs time curve for 7.5 m/s vel. with simply supported edges, SSSS

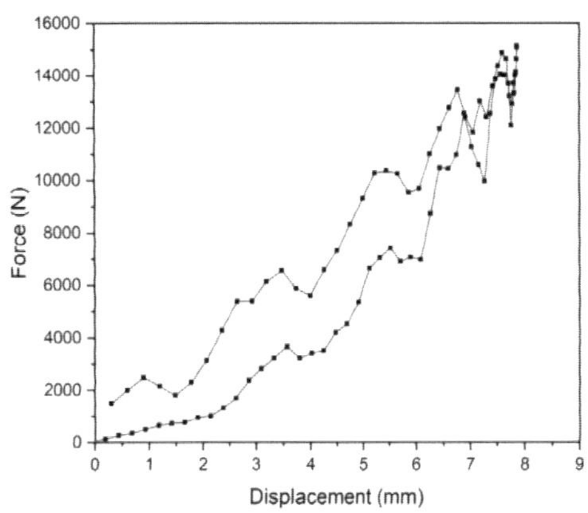

Figure 11: Impact force vs disp. curve for 7.5 m/s vel. with simply supported edges, SSSS

2.1.4 Impactor Velocity 10 m/s

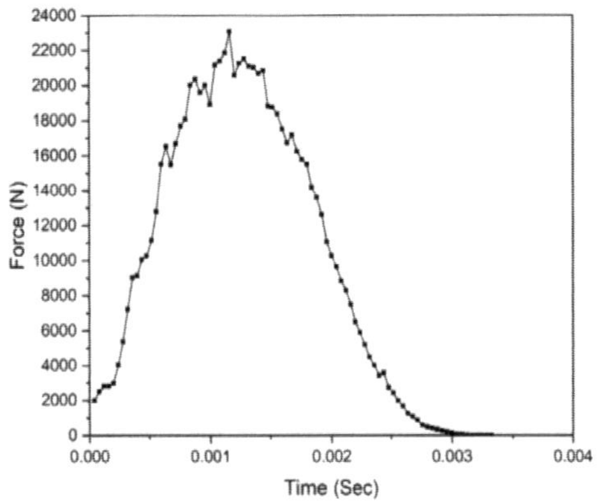

Figure12: Impact force vs time for 10 m/s vel. with clamped supported edges, CCCC

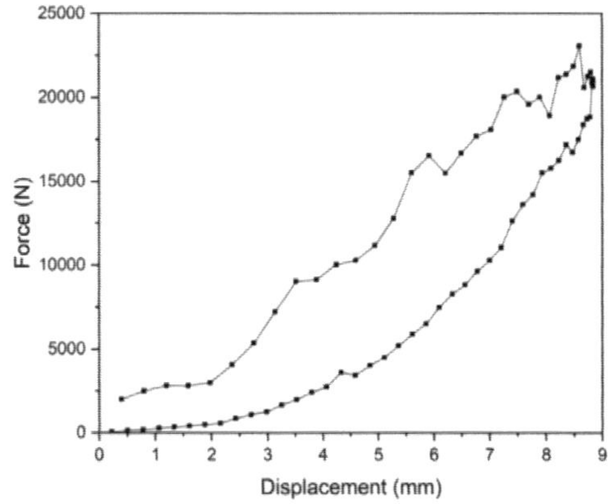

Figure 13: Impact force vs displacement for 10 m/s vel. with clamped supported edges, CCCC

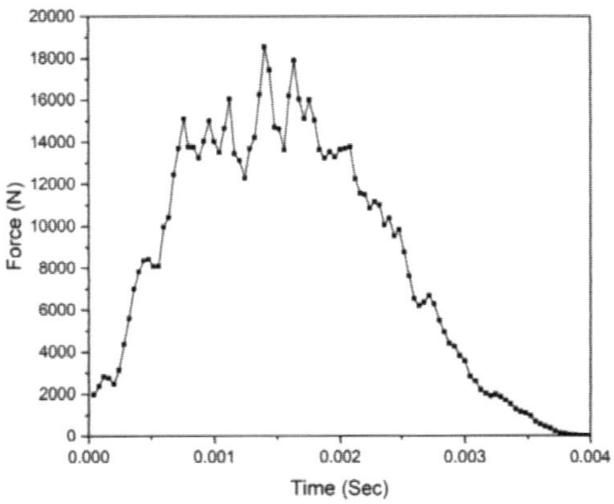

Figure 14: Impact force vs time curve for 10 m/s vel. with simply supported edges, SSSS

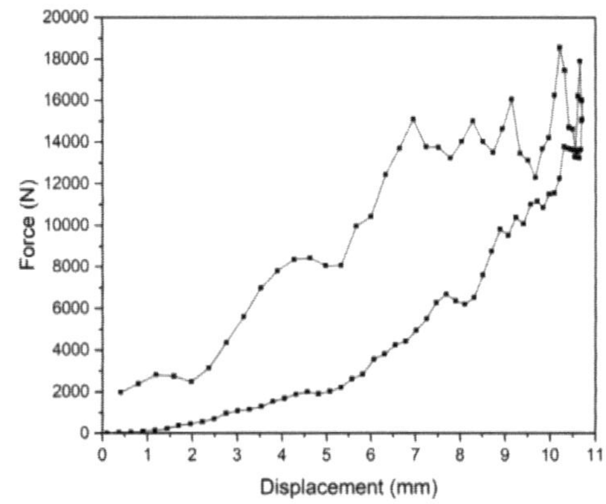

Figure15: Impact force vs displacement curve for 10 m/s vel. with[1]simply supported edges, SSSS

Table 1: Max. force and max. displacement for 2.5 to 10 m/s velocities.

BC ⟶		Clamped supported edges, CCCC	Simply supported edges, SSSS
Impactor velocity 2.5	Force (N)	4841.15	4464.47
	Displacement (mm)	2.52062	2.69725
Impactor velocity 5 m/s	Force (N)	10661.7	8714.63
	Displacement (mm)	4.83095	5.4
Impactor velocity 7.5	Force (N)	17014.01	15131.01
	Displacement (mm)	6.89633	7.87083
Impactor velocity 10	Force (N)	23067.86	18553.82
	Displacement (mm)	8.84218	10.70158

From the results following observation are seen

1. It is found that as impactor velocity increases the force and displacement also increases.

2. In CCCC boundary condition force is higher than SSSS boundary condition and displacement is lesser in CCCC.
3. The fluctuation in force time/ displacement curve is more in Simply supported edges SSSS may be due to fiber failure, matrix failure or other failure modes.
4. Failure is higher in impactor velocity10 m/s.

3. Validation of Numerical results with the experimental and numerical results from Liu et al. [1]

Drop weight impact is used in ABAQUS. Geometrical Modelling, meshing, material properties, boundary condition, interaction between impactor and laminates, between lamina to lamina properties, other parameters are established using ABAQUS/Explicit. The 25J impact energy is used for as specified in [1]. Dynamic penalty method is used between impactor and laminates. The friction contact coefficient is 0.3. Material T700GC/M21 is used for layup $[0_2/45_2/90_2/-45_2]_s$. The geometry size for laminate is 100mm×150mm×4mm and ply thickness is 0.25mm. The number of all elements is 60352. Hongkarnjanakul et al. [6] and Liu et al. [1] got 3.5ms impact contact time by experiments, and thus we considered 4.0ms impact time period for simulation. Numerical and experimental data taken from Liu et al. [1] using X-Y plotter software for comparison.

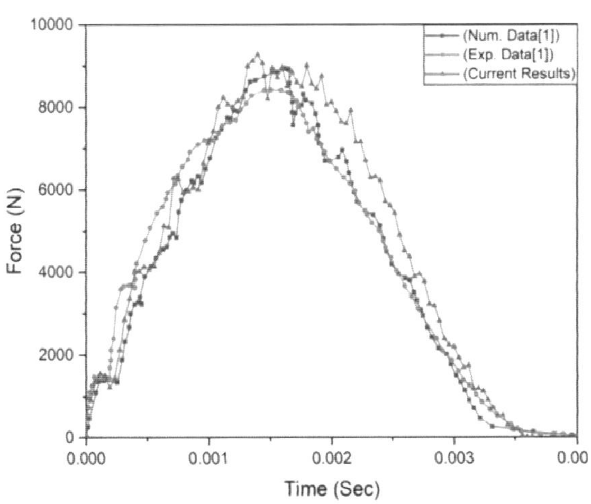

Figure16: Impact force vs time curve

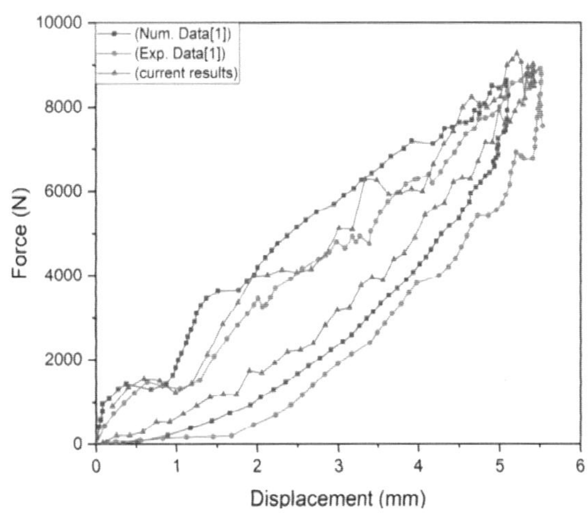

Figure17: Impact force vs displacement curve

Table 2: Comparison of current numerical result with reference paper [1] data.

	Numerical data from [1]	Experimental data from [1]	Current results
Force	8318.04 N at 0.0017 sec	8426.84 N at 0.0015 sec	8714.63 N at 0.0015 sec
Displacement	5.1 mm	5.5 mm	5.4 mm

Fig 16 and 17 shows the validation of current numerical results with the numerical and experimental results of P. F. Liu et al. [1]. The peak impact force for numerical and experimental taken from Liu et al. [1] are found to be 8318.04 N at 0.0017 sec and 8426.84 N at 0.0015 sec respectively. The impact for of current numerical simulation is 8714.63 at 0.00156 sec which is acceptable. The max. displacement for numerical and experimental taken from Liu et al. [1] are found to be 5.1 mm and 5.5 mm respectively. The displacement for current numerical simulation is 5.4 mm which very close to the existing results. The variation of results may be due to numerical filters which is used by [1]. Some numerical oscillations appear in the curve that may be because of numerical effects and dynamic responses. The impactor starts rebound at time 0.0017 sec approx, after the peak force. In the numerical simulation, Hasin's criterion is used, which is already inbuilt in ABAQUS package.

4. Conclusion:

This paper aims to discuss the effects of Impactor velocity and different boundary conditions on the force-time and displacement histories of CFRP laminates under low velocity impact. The impact responses are classified and discussed according to impactor velocity and boundary condition. Hashin failure criteria which is inbuilt in ABAQUS is very useful for developing solution. In numerical analysis, we can easily determine different failure like fiber failure, matrix failure etc. with help of force time/Displacement curve. There are other failure criteria like Chang-Chang and Puck criteria, which can be used for analysis.

The research can further be extended for this model like other validations such as the effect of different layup, stacking sequence, the effect of boundary condition. The effect of different friction coefficient on numerical results can be explored. Selection of different parameters are important for enhancing the impact strength.

Acknowledgment: Authors would like to acknowledge the financial support provided by Visvesvaraya National Institute of Technology, Nagpur to this research work.

References

[1] Liao B B and Liu P F 2017 Finite element analysis of dynamic progressive failure of plastic composite laminates under low velocity impact *Compos. Struct.,* **159,** 567–78.

[2] Karakuzu R and Okutan B 2003 The response of laminated composite plates under low-velocity impact loading *Compos. Struct.,* **59,** 119–27.

[3] Tita V, de Carvalho J and Vandepitte D 2008 Failure analysis of low velocity impact on thin composite laminates: Experimental and numerical approaches *Compos. Struct.,* **83,** 413–28.

[4] Feng D and Aymerich F 2014 Finite element modelling of damage induced by low-velocity impact on composite laminates *Compos. Struct.,* **108,** 161–71.

[5] Zhang J and Zhang X 2015 An efficient approach for predicting low-velocity impact force and damage in composite laminates *Compos. Struct.,* **130,** 85–94.

[6] Hongkarnjanakul N, Bouvet C and Rivallant S 2013 Validation of low velocity impact modelling on different stacking sequences of CFRP laminates and influence of fibre failure *Compos. Struct,* **106,** 549–559.

[7] Shi Y, Swait T and Soutis C 2012 Modelling damage evolution in composite laminates subjected to low velocity impact *Compos. Struct.,* **94,** 2902–13.

[8] Tan W, Falzon B G, Chiu L N S and Price M 2015 Predicting low velocity impact damage and Compression-After-Impact (CAI) behaviour of composite laminates *Compos. Part A Appl. Sci. Manuf.,* **71,** 212–26

[9] Wang S X, Wu L Z and Ma L 2010 Low-velocity impact and residual tensile strength analysis to carbon fiber composite laminates *Mater. Des.,* **31,** 118–25

[10] Hashin Z and Rotem A 1973 A Fatigue Failure Criterion for Fiber Reinforced Materials *J. Compos. Mater.,* **7,** 448–64

[11] Nisini E, Santulli C and Liverani A 2017 Mechanical and impact characterization of hybrid composite laminates with carbon, basalt and flax fibres *Compos. B.* Eng., **127,** 92–9

[12] Lopes C S, Seresta O, Coquet Y, Gürdal Z, Camanho P P and Thuis B 2009 Low-velocity impact damage on dispersed stacking sequence laminates. Part I: Experiments *Compos. Sci. Technol.,* **69,** 926–36

[13] Bienias J, Jakubczak P and Dadej K 2016 Low-velocity impact resistance of aluminium glass laminates - Experimental and numerical investigation *Compos. Struct.,* **152,** 339–48

[14] Mili F and Necib B 2001 Impact behavior of cross-ply laminated composite plates under low velocities*Compos. Struct.,* **51,** 237–44

[15] Wagih A, Maimí P, Blanco N, García-Rodríguez S M, Guillamet G, Issac R P, Turon A and Costa J 2019 Improving damage resistance and load capacity of thin-ply laminates using ply clustering and small mismatch angles *Compos. Part A Appl. Sci. Manuf.,* **117,** 76–91

[16] Buitrago B L, Santiuste C, Sánchez-Sáez S, Barbero E and Navarro C 2010 Modelling of composite sandwich structures with honeycomb core subjected to high-velocity impact *Compos. Struct.,* **92,** 2090–6

[17] Behnia S, Daghigh V, Nikbin K, Fereidoon A and Ghorbani J 2016 Influence of Stacking Sequence and Notch Angle on the Charpy Impact Behavior of Hybrid Composites *Mech. Compos. Mater.,* **52,** 489–96

IOP Conf. Series: Materials Science and Engineering 1004 (2020) 012018 doi:10.1088/1757-899X/1004/1/012018

[18] Hosur M V, Adbullah M and Jeelani S 2005 Studies on the low-velocity impact response of woven hybrid composites *Compos. Struct.*, **67,** 253–62
[19] Lopes C S, Camanho P P, Gürdal Z, Maimí P and González E V 2009 Low-velocity impact damage on dispersed stacking sequence laminates. Part II: Numerical simulations *Compos. Sci. Technol.*, **69,** 937–47
[20] Shi Y, Pinna C and Soutis C 2014 Modelling impact damage in composite laminates: A simulation of intra- and inter-laminar cracking *Compos. Struct.*, **114,** 10–9
[21] Bouvet C, Rivallant S and Barrau J J 2012 Low velocity impact modeling in composite laminates capturing permanent indentation *Compos. Sci. Technol.*, **72,** 1977–88
[22] Donadon M v., Iannucci L, Falzon B G, Hodgkinson J M and de Almeida S F M 2008 A progressive failure model for composite laminates subjected to low velocity impact damage *Comput. Struct.*, **86,** 1232– 52
[23] Ravandi M, Teo W S, Tran L Q N, Yong M S and Tay T E 2017 Low velocity impact performance of stitched flax/epoxy composite laminates *Compos. B.* Eng., **117,** 89–100
[24] Jefferson A J, Srinivasan S M and Arockiarajan A 2019 Effect of multiphase fiber system and stacking sequence on low-velocity impact and residual tensile behavior of glass/epoxy composite laminates *Polym. Compos.*, **40,** 1450–62
[25] Hashin Z 1981 Fatigue failure criteria for unidirectional fiber composites *J. Appl. Mech.*, **48,** 846–52

Bending analysis of a sandwich plate with elliptical/circular cutout

Achchhe Lal [1], B.M Sutaria[2] and Rahul Kumar[3]

[1] Assistant Professor, Department of Mechanical Engineering SVNIT, Surat, India.
[2] Associate Professor, Department of Mechanical Engineering SVNIT, Surat, India.
[3] Research Scholars, Department of Mechanical Engineering SVNIT, Surat.,.India.
*E-Mail: *lalachchhe@yahoo.co.in*

Abstract: Applications of sandwich plate as light weight structural member rapidly increased since last two decades because of its higher flexural rigidity to weight ratio. From aerospace vehicle to racing car, from giant submarine to small boat, from bicycle to high speed train all these weight sensitive areas have various parts made up by composite sandwich plates. Sometimes these members have hole of various shape for access of internal parts, lighting, ventilation etc. due to presence of these geometrical discontinuities in the form of hole, the structure gets more vulnerable to failure. During functioning, these parts usually acted upon by transverse force which may cause large deflection in the member or failure of the structure. Therefore, proper analysis of these structural member is desired for efficient and safe design. In the present work bending analysis of sandwich plate with elliptical and circular hole is carried out using finite element based first order shear deformation theory. A MATLAB program is developed for the present analysis. The efficiency of the program is found good as present results shows good agreements with the already published results.
Keywords: sandwich plate, elliptical/circular hole, bending

1. Introduction

Use of conventional material as structural member has been replaced by composite material very rapidly due its outstanding mechanical properties and light in weight in the weight sensitive area like aerospace vehicles, submarines, structural parts of racing cars, helicopter blades and many more. Further to increase the transverse rigidity sandwich plate is used in the place of composite plate which may further reduce the weight increased the transverse rigidity of structure. For, access of internal parts of structures, ventilation, lighting, fastening and sometimes to balance the system holes are provided to the structural member. Due to these holes the performance of the structure is directly influenced, therefore proper analysis of these structures is required for better safety and reliability.

Extensive studies are available for structural analysis of composite and sandwich plate with cutout under bending and finding their applications. An analysis for stresses and SCF was done by Paul and Rao [1,2] for laminated composite plate with circular hole under transverse loading. Jain [3] had analyzed plate with circular hole under transverse loading and revealed the variation of SCF with ratio (D/A) of diameter of hole (D) to plate length (A). In their work, they had examined the influence of parameters like hole size, plate thickness and nature of loading of SCF. XIWU et al. [4] had examined the variation of SCF with number of holes, distance between holes, and ply layups for a finite laminated composite plate with multiple elliptical holes. The effect of arrangement of cutouts along the length and width on stress concentration factor, maximum stress and deflection for the rectangular and square plates were investigated by Rzayyig [5]. An analysis for Stress around an arbitrary shape of hole in anisotropic plate under in-plane loading was done by Ukadgaonker and Rao [6] and had obtained results for various shapes of holes, nature of loadings and laminate geometry. SCF around a circular hole in a composite plate (carbon/epoxy) was experimentally determined by Toubal et al. [7] by using electronic speckle pattern interferometer (ESPI) and results were verified with theoretical model developed by Lekhnitskii. The effect of stacking sequence, reinforcement angles, angle of loading, materials, corner radii of the cutout on the failure strength and moment distribution in a laminated composite plate with square cutout were analysed by Patel and Sharma [8]. FEM with ABAQUS were considered to analyse composite plates with or without cutout acted upon axial, transverse and their combination of loading and the results were verified for ultimate load, maximum displacements and failure mode with experimental outcomes by Priyadharshani et al. [9]. FEM based on higher-order shear deformation theory was used to investigate the bending behavior of axisymmetric laminated shell under uniform pressure by Ram and Babu [10]. The same method was adopted for hyper shell stiffened along the margin of cutouts with different boundary conditions by Chaudhuri et al. [11]. Ghannadpour and Mehrparvar [12] had introduced a new technique to analyze influence of cutout shape, size and position for composite laminates with various boundary constraints. Zhou et al. [13] estimated damage index to improve accuracy for damage identification and verified the results with ultrasonic testing conducted on composite plate with cutout.

From the literature survey it is observed that lots of research work related to analysis of plate with geometrical discontinuities are available. To the best of author knowledge, no work to reveal the effect of cutout dimension on transverse deflection and nonuniform stress distribution are available. Finite element method is the one of the most efficient technics to solve complex geometry problem.

In the present work bending analysis of sandwich plate with elliptical and circular hole is carried out using finite element based first order shear deformation theory. Using a MATLAB program effect of multiplying factor, boundary conditions and cutout dimension of elliptical/circular hole dimension on deflection and distribution are investigation.

2. Mathematical Formulation

Here in the present work, finite element based first order shear deformation theory is considered for the bending analysis of sandwich plate with elliptical and circular hole with a MATLAB program. Four noded quadrature elements with five degree of freedom is considered. Material properties for skin of the sandwich plates is taken by multiplying a factor to that of core material. Properties of the core material can be given as,

$$Q_{core}=\begin{bmatrix} 0.999781 & 0.231192 & 0 & 0 & 0 \\ 0.231192 & 0.524886 & 0 & 0 & 0 \\ 0 & 0 & 0.262931 & 0 & 0 \\ 0 & 0 & 0 & 0.266810 & 0 \\ 0 & 0 & 0 & 0 & 0.159914 \end{bmatrix} \quad (1)$$

and material properties for top and bottom skin is taken as,

$$Q_{skin} = \text{Reinforcement factor} \times Q_{core} \quad (2)$$

Where the value of reinforcement factor can be chosen as factor to which we want to strengthen the top and bottom skin with respect to core material.

Plate considered in the present study has dimensions and subjected to uniform pressure. Non-dimensionalized deflection and stresses are evaluated along with their variations with dimension of hole. Major and minor axis diameter of elliptical hole denoted by 'ma' and 'mi' respectively.

2.1 Displacement field

First order shear deformation theory-based displacement field considered in the present study is,

$$\begin{aligned} u_1(x,y,z) &= u(x,y) + z\Phi_x(x,y) \\ u_2(x,y,z) &= v(x,y) + z\Phi_y(x,y) \\ u_3(x,y,z) &= w(x,y) \end{aligned} \quad (3)$$

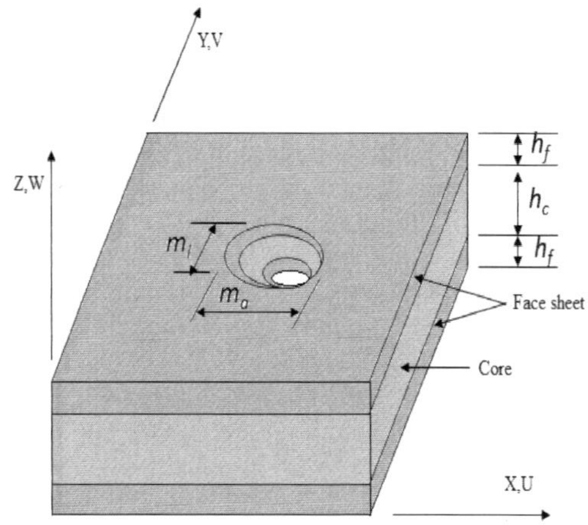

Figure 1: Laminate sandwich plate with cutout

2.2. Strains

Expression for strains are obtained by differentiating the displacement functions as,

$$\begin{Bmatrix} \varepsilon_x \\ \varepsilon_y \\ \gamma_{xy} \\ \gamma_{yz} \\ \gamma_{zx} \end{Bmatrix} = \begin{bmatrix} \frac{\partial u_1}{\partial x} \\ \frac{\partial u_2}{\partial x} \\ \frac{\partial u_2}{\partial x} + \frac{\partial u_1}{\partial y} \\ \frac{\partial u_2}{\partial z} + \frac{\partial u_3}{\partial y} \\ \frac{\partial u_3}{\partial x} + \frac{\partial u_1}{\partial z} \end{bmatrix} \quad (4)$$

2.3 Kinematic matrix

The corresponding kinematic matrix is written as,

$$[B] = \begin{bmatrix} \frac{\partial N}{\partial x} & 0 & 0 & 0 & 0 \\ 0 & \frac{\partial N}{\partial y} & 0 & 0 & 0 \\ \frac{\partial N}{\partial y} & \frac{\partial N}{\partial x} & 0 & 0 & 0 \\ 0 & 0 & 0 & \frac{\partial N}{\partial x} & 0 \\ 0 & 0 & 0 & 0 & \frac{\partial N}{\partial y} \\ 0 & 0 & \frac{\partial N}{\partial x} & N & 0 \\ 0 & 0 & \frac{\partial N}{\partial y} & 0 & N \end{bmatrix} \quad (5)$$

Where, N is shape function.

2.4 Stresses from strain

$$\begin{bmatrix} \sigma_x \\ \sigma_y \\ \sigma_{xy} \\ \tau_{yz} \\ \tau_{xz} \end{bmatrix} = \left[\bar{Q}\right] \begin{bmatrix} \varepsilon_x \\ \varepsilon_y \\ \varepsilon_{xy} \\ \gamma_{yz} \\ \gamma_{xz} \end{bmatrix} \tag{6}$$

Where, $\left[\bar{Q}\right]$ is reduced stiffness matrix.

2.5 Stiffness matrix

$$K = [K_l] + [K_{nl}] \tag{7}$$

where $[K_l] = \sum_{k=1}^{nL} \int_A B^T D\, B \; dz\, dA$ (8)

is linear stiffness matrix and K_{nl} is nonlinear stiffness matrix given as,

$$\begin{aligned} K_{nl} = & \frac{1}{2}\sum_{k=1}^{nL}\int_A B^T D_3\{A\}\{\phi\}dA + \\ & \frac{1}{2}\sum_{k=1}^{nL}\int_A \{A\}^T\{\phi\}^T D_4 \mathrm{B} dA + \\ & \frac{1}{2}\sum_{k=1}^{nL}\int_A \{\phi\}^T\{A\}^T D_5\{\mathrm{A}\}\{\phi\}\, dA \end{aligned} \tag{9}$$

2.6. Governing equation for bending

The equation of governing bending problem ca be written as,

$$[K]\{\mathrm{X}\} = [P] \tag{10}$$

where $\{\mathrm{X}\} = [u_0 \quad v_0 \quad w_0 \quad \Phi_y \quad \Phi_x]$ is displacement vector and $[P]$ is load vector.

3. Results and Discussions

Transverse central deflection and stress are evaluated for sandwich plate with and without elliptical/circular hole under uniform pressure using MATLAB program in the present work. Influence of hole dimensions and reinforcement factor on transverse central deflection and stresses are examined. Table 1 shows validation of the present results along with the variation of transverse central deflection and stresses with hole dimension. It is observed from the Table 1 that with increase in hole dimension the plate gets weaken and stresses increases because of decrement in the overall resisting area.
As major axis diameter of elliptical hole increases from 0.2 to 0.4 and 0.6 transverse central deflection increases from 1.56% to 4% and 11.3% with respect to that of plate without hole. Plate with circular hole shows 35.18% deflection as compared to that of plate without hole.

Table 2 reveals the effect of reinforcement factor on transverse central deflection along with the effect of hole dimension. With the increment of reinforcement factor the plate gets strengthen due to this the transverse central deflection decreases. As reinforcement factor raises from 5 to 10, transverse central deflection decreases 38 % and this decrement gets 23.5% as reinforcement factor rises to 15. The effect of elliptical hole dimension on transverse central deflection is also revealed by this Table 2.

Table 1: Effect of cutout dimension on NTMD and stress for bending of three layered (0/C/0) sandwich plate.

Cutouts dimension	W_0 $(a/2,b/2,0)$	σ_x^1 (σ_y^1) $(a/2,b/2,-h/2)$	σ_x^2 (σ_y^2) $(a/2,b/2,-2h/5)$	σ_x^3 (σ_y^3) $(a/2,b/2,-2h/5)$	τ_{xz}^1 $(0,a/2,0)$	τ_{yz}^1 $(0,a/2,0)$
Present (without cutout)	257.18	59.105 (38.241)	47.284 (30.432)	9.456 (6.094)	2.999	3.997
Ferreira el al. [14]	257.52	59.967 (38.320)	46.291 (29.974)	9.258 (5.995)	2.390	----
m_a=0.2 m_i=0.1	261.21	59.360 (38.131)	47.488 (30.505)	9.497 (6.101)	3.135	3.677
m_a=0.4 m_i=0.1	267.68	60.341 (37.908)	48.273 (30.327)	9.654 (6.065)	3.048	3.243
m_a=0.6 m_i=0.1	286.37	66.279 (36.204)	53.023 (28.963)	10.604 (5.792)	3.001	3.005
m_a=0.1 m_i=0.2	261.25	60.278 (38.357)	48.223 (30.685)	9.644 (6.137)	3.877	3.387
m_a=0.1 m_i=0.4	272.48	61.913 (40.830)	49.531 (32.664)	9.906 (6.532)	3.298	3.491
m_a=0.1 m_i=0.6	314.29	58.611 (53.505)	46.889 (42.804)	9.377 (8.560)	3.142	3.713
m_a=0.5 m_i=0.5	347.68	67.467 (45.518)	53.974 (36.414)	10.794 (7.282)	3.317	3.589

Table 2: Effect of multiplying factor and cutout dimension on maximum transverse deflection.

Cutouts dimension		W_0				
	R=5	**R= 10**	**R= 15**	**R=20**	**R=25**	**R=30**
m_a=0.2 m_i=0.1	261.26	161.28	123.38	103.36	90.94	82.45
m_a=0.4 m_i=0.1	267.74	165.88	127.23	106.80	94.133	85.483
m_a=0.6 m_i=0.1	286.43	178.64	137.69	116.04	102.61	93.454
m_a=0.1 m_i=0.2	261.30	161.47	123.62	103.61	91.205	82.724
m_a=0.1 m_i=0.4	272.54	168.81	129.37	108.48	95.492	86.599
m_a=0.1 m_i=0.6	272.54	195.51	149.93	125.74	110.55	100.08
m_a=0.5 m_i=0.5	347.76	216.13	165.61	138.67	121.85	110.28
without cutout	257.24	159.01	121.81	102.17	89.993	81.679

Table 3: Influence of support conditions on NTMD of square sandwich (0/C/0) plate with circular hole of diameter 0.1 and R=5 subjected to uniform transverse loading.

Cutouts dimension	W_0 $(a/2, b/2, 0)$			
	SSSS	**CCCC**	**CSCS**	**CCSS**
ma=0.2 mi=0.1	261.26	111.74	159.40	166.75
ma=0.4 mi=0.1	267.74	115.55	170.62	171.18
ma=0.6 mi=0.1	286.43	126.43	199.24	181.66
ma=0.1 mi=0.2	261.30	112.07	159.00	167.56
ma=0.1 mi=0.4	272.54	116.92	160.92	169.96
ma=0.1 mi=0.6	272.54	135.46	171.75	193.30
ma=0.5 mi=0.5	347.76	152.99	214.76	201.10
without cutout	257.24	104.38	151.41	162.82

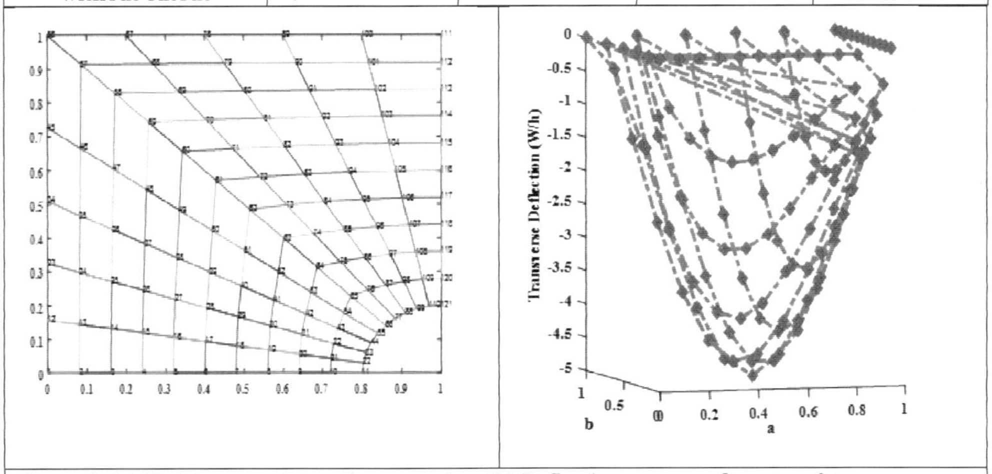

Figure 2: (a) Nodal connectivity of quarter plate (b) Deflection contour of quarter plate

Table 3. shows influence of support conditions on NTMD of sandwich plate with elliptical hole of different size, when plate is subjected to uniform pressure. With increment of constraint on edge of plate, the NTMD decreases. There is maximum NTMD when all edges of plate are simply supported and least for all edges clamped. The

NTMD decreases by 57.23 % when the support condition changes from simply supported (SSSS) to clamped condition (CCCC). The plate shows more stability when alternate edges are clamped and simply supported than two adjacent edges are clamped and other two are simply supported due to symmetricity in former case.

4. Conclusions

In the present paper transverse central deflection and inter laminar stresses are evaluated for a sandwich plate subjected to uniform pressure. Effect of strengthening factor of top and bottom skin on the above two parameters are also examined. Some important conclusions made from
the present study can be listed as,

- With increment of reinforcement factor flexural rigidity of the plate gets improved.
- Plate without any hole is stronger than the plate with hole.
- As hole diameter increases, plates get more and more weak i.e. transverse central deflection increases.

References

[1] Paul T K, Rao K M 1989 Stress analysis around circular holes in FRP laminates under transverse load *Computers & Structures*, **33**, 923-937.

[2] Paul T K, Rao K M 1993 Finite element evaluation of stress concentration factor of thick laminated plates under transverse loading *Computers & Structures,* **48**, 311-31.

[3] Jain N K 2009 Analysis of stress concertation and deflection in isotropic and orthotropic rectangular plates with maximum circular hole under transverse loading *International Journal of Mechanical and Mechatronics Engg,* **3,** 1513-1519.

[4] Xiwu X, Liandxing S, Xuqi F 1995 Stress concentration of finite composite laminates weakened by multiple elliptical holes *Int. J. Solid Structures,* **32,** 3001-3014.

[5] Rzayyig AY 2011 Effect of cutouts on the behaviour of clamped rectangular plates, Anbar *Journal for Engineering* Sciences, **2,** 45-59.

[6] Ukadgaonker V G, Rao D K N 2000 A general solution for stresses around holes in symmetric laminates under inplane loading *Composite Structures,* **49,** 339-354.

[7] Toubal L, Karama M, Lorrain B 2005 Stress concentration in a circular hole in composite plate *Composite Structures,* **68,** 31-36.

[8] Patel N P, Sharma D S 2015 Bending of composite plate weakened by square hole *International Journal of Mechanical Sciences,* **94,** 131-139.

[9] Priyadharshani S A, Prasad A M, Sundaravadivelu R 2017 Analysis of GFRP stiffened composite plates with rectangular cutout *Composite Structures,* **169,** 42-51.

[10] Sai Ram K S, Babu T S 2001 Study of bending of laminated composite shells. Part II: shells with a cutout *Composite Structures,* **51,** 117-126.

[11]Chaudhuri P B, Mitra A, Sahoo S 2017 Deflection, forces and moments of composite stiffened hyper shells with cutout *Materials Today*: Proceedings, **4,** 9718-9722.

[12] Ghannadpour S A M, Mehrparvar M 2018 Energy effect removal technique to model circular/elliptical holes in relatively thick composite plates under in-plane compressive load *Compos Struct* , **202,** 1032-1041.

[13] Zhou J, Li Z, Chen J 2018 Damage identification method based on continuous wavelet transform and mode shapes for composite laminates with cutouts *Composite Structures,* **191,** 12-23.

[14] Ferreira A J M 2005 Analysis of Composite Plates Using a layerwise theory and multiquadric discretization *Mechanics of Advanced Materials and Structures,* **12,** 99-112.

Post-buckling analysis of the shape memory polymer sandwich composite beam under dynamic temperature variation

Achchhe Lal[1*] and Kanif Markad[2]
[1,2] Department of Mechanical Engineering, S.V.N.I.T., Surat-395007, Gujarat, India.
[*]E-Mail: lalachchhe@yahoo.co.in

Abstract: Post-buckling analysis of the shape memory polymer sandwich composite (SMPSC) beam under HSDT utilizing von Karman kinematics using FEM is performed in present analysis. The aim of the study is to develop a model which accurately perform the buckling analysis. The nondimensional critical buckling load (NCBL) evaluation under the action of inplane uniform load with different boundary conditions (BC), plate thickness ratio, under dynamic temperature variation for SMPSC. The study clearly reveled the differentiation between SMPC and SMPSC beam.
Keyword: HSDT, SMPSC, FEM, NCBL, glass transition temperature (T_g)

1. Introduction

Shape memory polymers (SMP) responds in the vicinity of various stimuli such as mechanical stress, heat, pH, moisture, light, electricity, magnetic field or solvents. It can be deformed into intermediate shape and as it comes in contact with corresponding sensation and regain its programmed shape. Temperature stimulated shape memory polymers have become popular in last three decades as an effective alternate of another widely used smart material - Shape Memory Alloy (SMA). Kumpfer et al. [1] utilized soft poly and hard metal phase for investigating shape fixing and recovery of materials. Considerable research has been done on the experimental estimation of glass transition region of SMP selected for analysis. Properties needed to perform analysis can either be obtained through experimentation or by curve-fitting of the experimental results Wang et al. [2]. Qi et al. [3] analyzed and performed the finite deformation thermo-mechanical behavior of thermally induced shape memory polymers. Lal and Markad [4, 5] performed the post buckling, deflection and stress analysis of MWCNT reinforced composite beam under HSDT utilizing FEM. Kumar and Srinivas [6] were examined the buckling investigation of functionally graded (FG) CNT beam, and also property of the composites were found either by Mori Tanaka or rule of mixture method. Many of them also analyzed the effect of beam aspect ratio, volume fraction of CNT (Vcnt) over buckling response of the beam. Karamanli and Aydogdu [7] performed the buckling analysis of composite laminated beam with the inplane variable load with different BC and beam aspect ratio. The properties of composite material is estimated through the theory of volume averaging, as per the assumption that new two phase matrix and fiber have similar deformations Shen et al. [8]. Zhang et al. [9] studied micro-buckling of fibres reinforced in Elastomeric Memory Composites (EMC) under deformation due to bending. Carbon reinforced in SMP matrix has been considered as EMC, applicable for deployable antennas subjected to high strains. Buckling phenomenon of EMC was elaborated and its calculated results were compared with Lan et al. [10], which indicated appreciable convergence to define accuracy of results.

From the studied limited literature it is observed that, no literature is observed who study the effect of post buckling analysis of shape memory polymer composite and sandwich structures under inplane UDL with HSDT utilizing FEM. In present paper, effect of dynamic temperature variation over post buckling analysis of SMPC and SMPSCB is analyzed for different thickness ratio, BC. Study also clearly defines the effect of the core thickness variation over buckling analysis of the composite beam.

2. Mathematical formulation

2.1 Modeling of SMP effect

Material properties of the selected SMP matrix has been extensively evaluated in the prior work of Kumpfer et al. [1]. Thus the characteristics required for the numerical analysis can either be experimentally investigated or can be determined through curve fitting of experimental data Wang et al. [2]. Storage modulus E_m (T) of polymers is find by Eq. (1) for wide range of temperatures across T_g with a constant frequency, according to X. Wang et al. [2].

$$E_m(T) = (E_2 - E_3).\exp\left(-\left(\frac{T}{Tg}\right)^{m_2}\right) + E_3 \tag{1}$$

The values for Eq. (1) unknown considered as $T_g = 68$ °C, E_2=1053.41 MPa, E_3=44.11 MPa and $m_2 = 41$. Poissons ratio is also temperature dependent property which can be calculated through phase transition model as per (Qi et al. 2008),

$$\mu_m = \mu_g f_g + \mu_r(1 - f_g) \text{ where, } f_g = 1 - \frac{1}{1 + \exp[-(T - T_m)/Z]} \tag{2}$$

Where reference temperature defined by T_m and f_g is the frozen phase volume fraction

2.2 Estimation of carbon fiber reinforced SMPC properties

Based on Eqs. (3-7), properties of laminated SMPC has been calculated and applied in the subsequent sections.

$$E_{c1} = E_{f1}v_f + E_m v_m \tag{3}$$

$$E_{c2} = (1 - C)E_{c2}^1 + CE_{c2}^2 \tag{4}$$

$$\mu_{c21} = (1 - C)\mu_{c21}^1 + C\mu_{c21}^2 \tag{5}$$

$$\mu_{c12} = \mu_{c21}\frac{E_{c2}}{E_{c1}} \tag{6}$$

$$G_{c12} = (1 - C)G_{c12}^1 + CG_{c12}^2 \tag{7}$$

Where E_{c1} and E_{c2} are modulus of composites along longitudinal and transverse directions respectively. E_{f1} and E_{f2} are Young's modulus of fiber along longitudinal and transverse directions respectively. G_{f12} is shear modulus along plane 1-2. G_{c12} is shear modulus of composite in 1-2 plane. C is transverse contact coefficient between fibers. μ_f is Poisson's ratio of carbon fiber. μ_{c12} and μ_{c21} indicate Poisson's ratio of composites along 1-2 and 2-1 plane respectively.

Table 1: Material properties of reinforcement [8]

Parameters	E_{f1}	E_{f2}	G_{f12}	μ_f	C
Values	2.3 × 10^5MPa	8.2 × 10^3MPa	2.73 × 10^4MPa	0.25	0.2

2.3 Displacement field model

In present analysis one dimensional composite beam is utilized for the buckling and vibrational analysis under variable inplane nonuniform loading. The composite beam is having length, width and thickness is *a*, *b* and *t* respectively, as shown in the figure 2.

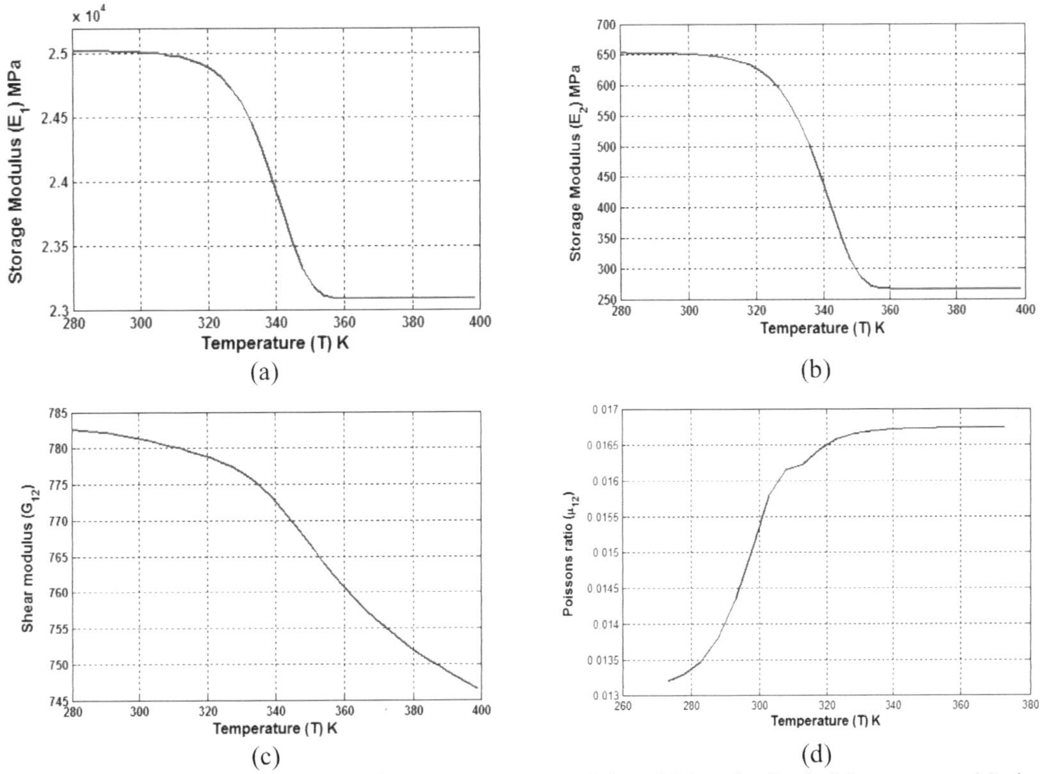

Figure 1: Influence of variation of temperature on modulus: (a) longitudinal, (b) transverse (c) shear modulus and (d) Poissons ratio of SMPC

$$\bar{u}(x,z) = \boldsymbol{u_0} + (z - \frac{4}{3}h^2 z^3)\Psi_x + (-C_2 Z^3)\emptyset_x;\ \bar{w}(x,z) = \boldsymbol{w_0} \tag{8}$$

where u_0, w_0, and $\emptyset = \partial w/\partial x$ are displacement along mid-plane axis, displacement along transverse direction, rotation of normal to the mid-plane along 'y'- axis and slope along 'x'- axis, respectively. Displacement vector for the modified C^0 continuous model can be expressed as,

$$\{q\} = [u \ \ w \ \ \emptyset_x \ \Psi_x]^T \tag{9}$$

2.4 Stress, Strain and displacement relation

Here, total strain is the combination of linear $\{\bar{\boldsymbol{\varepsilon}}^{\boldsymbol{L}}\}$ and nonlinear strain vector $\{\bar{\boldsymbol{\varepsilon}}^{\boldsymbol{NL}}\}$ with von Karman nonlinearity expressed as, Lal and Markad [4, 5],

$$\{\bar{\boldsymbol{\varepsilon}}\} = [B]\{\boldsymbol{q}\} + \frac{1}{2}[A_{nl}]\{\boldsymbol{\emptyset_{nl}}\} \tag{10}$$

Similarly, plane stress and total strain relationship expressed as, Lal and Markad [4, 5],

$$\{\bar{\sigma}\} = [\overline{Q_{ij}}]\{\bar{\varepsilon}\} \tag{11}$$

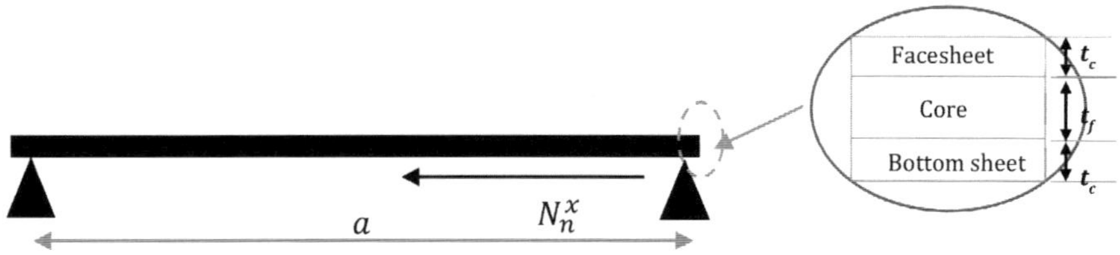

Figure 2: Geometrical configuration of sandwich composite beam

2.5 Finite element analysis of the composite beam

In the present work, a C^0 1-D Hermitian beam element with four degrees of freedom per node is employed, [Lal and Markad 2018], rewritten as this type of beam element geometry and the displacement vector.

$$\{\boldsymbol{q}\} = \sum_{i=1}^{NN} N_i \{\boldsymbol{q}\}_i\,; \qquad x = \sum_{i=1}^{NN} N_i\, x_i; \tag{12}$$

In order to calculate transverse displacement and slope, linear interpolation for axial displacement and rotation of normal and Hermite cubic interpolation functions are considered.

$$\Pi_1 = \sum_{e=1}^{NE} \Pi_a^{(e)} = \sum_{e=1}^{NE} \left(\boldsymbol{U}_L^{(e)} + \boldsymbol{U}_{NL}^{(e)}\right) \tag{13}$$

Where, NE and (e) denote the number of elements and elemental, respectively.
Eq. (13) can be further evaluated as,

$$\Pi_1 = \frac{1}{2}\sum_{e=1}^{NE} \left[\{\boldsymbol{q}\}^{T(e)}[K_l + K_{nl}]^e\{\boldsymbol{q}\}^{(e)}\right] = \{\boldsymbol{q}\}^T[K_l + K_{nl}]\{\boldsymbol{q}\} \tag{14}$$

Work done due to inplane variable nonuniform loading can be represented as follows, [5]

$$\Pi_2 = \frac{1}{2}\int_A N_n^x \left\{\frac{\partial u}{\partial x}\right\}^2 dA \text{ where, } N_n^x = N \tag{15}$$

N is the inplane uniformly distributed load (UDL) as shown in the figure 3.

Mechanically induced buckling and vibrational analysis of the inplane variable hydrostatic loading over laminated composite beam evaluated by the minimization of first variation of (Π_1-Π_2) with respect to generalized displacement vector is given by, Lal and Markad [4, 5]

$$\frac{\partial}{\partial\{q\}}(\Pi_1 - \Pi_2) = 0 \tag{16}$$

$$\{[K_L + K_{NL}] - \lambda[K_G]\}\{q\} = 0$$

$$[K]\{q\} = \lambda[K_G]\{q\} \tag{17}$$

$$[K]\{q\} = \lambda[M]\{q\} \tag{18}$$

$$[K\ ] = \{[K_l] + [K_{nl}] - [K_G]\}$$

Where, $[K_G]$ is the geometric stiffness matrix, $[M]$ is the global consistent mass matrix [27] and $\{q\}$ is the transverse deflection. The solution of the above buckling standard equation obtained by Newton Raphson method which is the most popular solution method due to its quick convergence. Eq. (17) and Eq. (18) corresponds to buckling and vibration analysis of the considered structure.

3. Results and discussion

User defined MATLAB program is developed based on defined formulation to find the post buckling of the SMP sandwich composite beam under dynamic temperature variation. One dimensional composite beam under uniformly distributed load is considered for the present simulation. Temperature dependent polymer is considered and its effect over analysis is studied over a wide range of temperature that includes glass transition region of SMPSC so that the buckling phenomenon with respect to uniform variation of temperature can be understood.

Table 2 shows the convergence study to select the proper number of element for further investigation without diluting the solutions. For the convergence analysis laminated composite beam is considered with ply orientation of [0/90/0], beam aspect ratio of 5 and
$E_1/E_2=40,\ E_2=E_3,\ G_{12}=G_{13}=0.5E_2,\ G_{23}=0.2E_2,\ \nu_{12}=\nu_{13}=\nu_{23}=0.25, a/h=5$ under inplane UDL. For the further analysis 30 number of element is considered as from neL=20 it starts to converge.

Table 2: Convergence of nondimensional buckling load of composite beam under various boundary conditions

a/h	Number of element (neL)	CF	CH	HH	CC
5	16	4.24	9.143	8.60	10.90
	20	4.24	9.142	8.51	10.81
	24	4.24	9.142	8.45	10.76
	30	4.24	9.142	8.40	10.72
10	16	5.81	24.53	18.95	34.55
	20	5.81	24.44	18.55	34.06
	24	5.80	24.37	18.28	33.77
	30	5.80	24.30	18.10	33.50

The compatibility and accuracy of the utilized analytical model is verified by validating it with Ref. [7] for laminated composite and sandwich beam under inplane UDL considering different number of layups in Table 3. For the validation purpose beam aspect ratio considered as 5 and properties of facesheet are, $E_1/E_2=25,\ E_2=E_3,\ G_{12}=G_{13}=0.5E_2,\ G_{23}=0.2E_2,\ \nu_{12}=\nu_{13}=\nu_{23}=0.25$ and core properties are
$m_{2c}=1.5; m_{3c}=0.4; E_{2c}=10.3e^9; \nu_{12c}=0.25; E_{1c}=m_{1c}*E_{2c}; G_{12c}=m_{2c}*E_{2c};$

$$m_{1c}=1;\ G_{13c}=G_{12c}; G_{23c}=m_{3c}*E_{2c}; \nu_{21c}=\nu_{12c}*E_{2c}/E_{1c}; \rho_c=1;\ NCBL=\frac{N_{cr}a^2}{E_1bh^3}$$

Table 3: Validation of nondimensional buckling load under various boundary conditions

Reference	No. of layer	CC	SS	CS	CF
Ref. [7] HBT	[0/90/0]	5.395	3.756	5.338	3.390
Present		5.982	4.431	4.941	2.224
Ref. [7] HBT	[0/90]	4.571	2.224	4.289	1.255
Present		4.917	3.212	3.815	1.360
Ref. [7] HBT	[0/C/0]	6.150	4.366	6.118	3.594
Present		6.576	4.901	5.506	2.355

Figure 3 shows the variation occurred in nondimensional critical load (NCBL) with respect to beam thickness ratio under different BC. Analysis is carried out at T_g for the sandwich composite beam [0/C/0] whose facesheet and bottom sheet made from SMPC. From the figure it is observed that, maximum NCBL observed with clamped BC whereas minimum is at CF boundary condition. With increasing thickness ratio, NCBL also get raised because of reducing overall thickness of the sandwich composite beam.

Figure 4 shows the effect of dynamic temperature variation in facesheet, bottom sheet in sandwich and SMPC over NCBL at thickness ratio of 5 with ply orientation of [0/C/0] under clamped boundary condition. Two important conclusions are drawn from this study. The very first one is the differentiation between sandwich and layered structure and second one is the effect of the core thickness over NCBL of sandwich structure. As compared to layered SMPC, maximum buckling load is observed in sandwich structure and is still increases with core thickness.

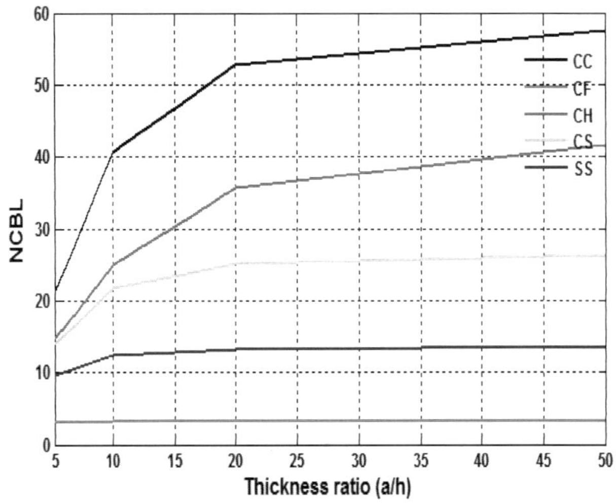

Figure 3: Effect of thickness ratio and BC over NCBL

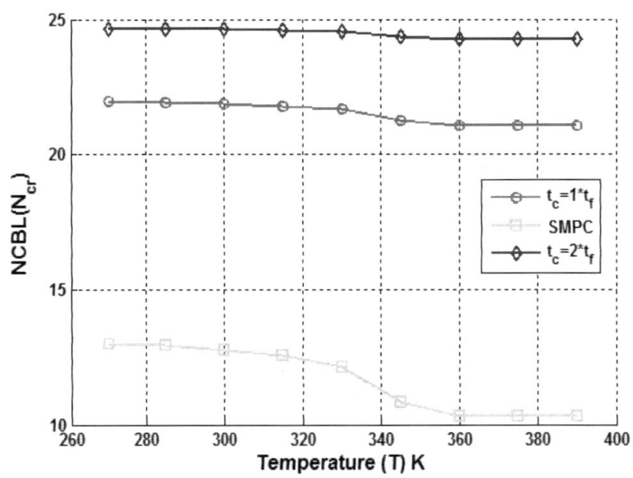

Figure 4: Effect of dynamic temperature variation over NCBL

4. Conclusion

The present study deals with the buckling analysis of 1-D SMPC and SMPSC beam through a simplified C^o continuity FEM based on HSDT. This work highlights the effect of dynamic temperature variation along with glass transition temperature on the NCBL of beam subjected to uniformly distributed inplane loading. Analysis identified the effect of beam thickness ratio, BC and core thickness variation over NCBL of sandwich composite beam.

References

[1] Kumpfer J R, Rowan S J 2011 Thermo-, photo-, and chemo-responsive shape-memory properties from photo-cross-linked metallo-supramolecular polymers. *J. Am. Chem. Soc.*, **133**, 12866–12874.
[2] Wang X, Lu H, Shi X, Yu K, Fu Y Q 2019 A thermomechanical model of multi-shape memory effect for amorphous polymer with tunable segment compositions *Composites Part B: Engineering*, **160**, 298–305. doi:10.1016/j.compositesb.2018.10.048
[3] Qi H J, Nguyen T D, Castro F, Yakacki C M, Shandas R 2008 Finite deformation thermo-mechanical behavior of thermally induced shape memory polymers *Journal of the Mechanics and Physics of Solids*, **56**, 1730–1751. doi:10.1016/j.jmps.2007.12.002
[4] Lal A, Markad K 2018 Deflection and stress behaviour of multi-walled carbon nanotube reinforced laminated composite beams *Computers and Concrete*, **22**, 501–514.
[5] Lal A, Markad K 2019 Thermo-Mechanical Post Buckling Analysis of Multiwall Carbon Nanotube-Reinforced Composite Laminated Beam under Elastic Foundation *Curved and Layer. Struct.*, **6**, 212–228.
[6] Kumar P, Srinivas J 2017 Vibration, buckling and bending behavior of functionally graded multi-walled carbon nanotube reinforced polymer composite plates using the layer-wise formulation *Composite structures*, **177**, 158–170.
[7] Karamanli A 2017 Bending analysis of composite and sandwich beams using Ritz method *Anadolu university journal of science and technology A-Applied sciences and engineering*, **19** (1), 10-23.
[8] Shen G L, Hu G, Liu B 2006 *Mechanics of composite materials -Science and Technology*, Beijing, China.
[9] Zhang F, Zhou T, Liu, Y, Leng J 2015 Microwave synthesis and actuation of shape memory polycaprolactone foams with high speed *Scientific Reports*, **5**, 11152. doi:10.1038/srep11152
[10] Lan X, Liu L, Liu Y, Leng J, Du S 2014 Post microbuckling mechanics of fibre-reinforced shape-memory polymers undergoing flexure deformation *Mechanics of Materials*, **72**, 46–60.

Fracture analysis of center crack functionally graded material plate under tensile biaxial loading by extended finite element method

Achchhe Lal [1*], B.M Sutaria[2] and Kundan Mishra[3]

[1] Assistant Professor, Department of Mechanical Engineering SVNIT, Surat, India.
[2] Associate Professor, Department of Mechanical Engineering SVNIT, Surat, India.
[3] Research Scholars, Department of Mechanical Engineering SVNIT, Surat., India.
*Email: lalachchhe@yahoo.co.in

Abstract: The fracture analysis of a centrally cracked functionally graded material (FGM) plate under biaxial stresses using extended finite element method (XFEM) is studied. In this present study the effect of crack length and crack angle on mixed mode stress intensity factor (MSIF) of center cracked FGM plate under biaxial stress is presented. The mathematical formulation is carried out in MATLAB by XFEM using interaction integral and partition of unity method. The results obtained by this present approach are compared with the result from the literature. This study is carried out to find MMSIF of center cracked FGM plate to understand the behavior of crack under biaxial state of stress.
Keywords: FGM plate, extended finite element method, MMSIF.

1. Introduction

Functionally graded material (FGM) is one of the emerging advanced composite where the material properties vary in desired direction. This concept was first utilized by Japan while working on a project where heat resisting material was required. Now FGM materials are being used for many engineering applications where high thermal resistance and high resistance to crack propagation are required.

In spite of sophisticated and advanced manufacturing technology, components contain cracks, which become very dangerous as components may fail under small loadings. So, researchers are giving more effort in analyzing the behavior of cracked plate under different loading conditions. In FGM material properties can be tailored in such a way that crack propagation can be minimized. In this direction Kim and Paulino [1] presented interaction Integral (M-integral) method to calculate MMSIF of orthotic FGM plate with straight or curved crack under loading. The proposed work is accurately compared with available analytical and numerical solutions. Bayesteh and Mohammadi [2] Studied fracture mechanics by XFEM, where singular stress field is reproduced by using orthotropic crack tip enrichments. In the present work it is verified that orthotropic XFEM needs lesser DOFs as compared to conventional FEM. Lal and Palekar [3] presented the stochastic MMSIF of cracked composite plate with random material properties. The proposed work is done by using XFEM and the present model is seen to be very accurate for uncertainty analysis. Mohammadi [4] presented fracture analysis by XFEM, where MMSIF of cracked FGM plate by using interaction integral and partition of unity method are discussed. Ebrahimi et al [5] investigated and modeled crack in orthotropic composite plate under uniaxial and bi-axial load by XFEM where re-meshing is not required while progress of crack. In the present study it is observed that crack tip domain size has not any effect on SIF when it reaches one third of crack length. Lal et al [6] investigated stochastic fracture analysis cracked composite beam under different loadings. The sensitivity of individual parameter on MMSIF is also investigated and here it is also confirmed that SIF is get reduced when orthotropic angle is above 40^0. So it is essential to keep orthotropic angle more than 40^0 for safety point of view.

Meek and Ainsworth [7] presented finite element analysis with lower and upper bond theory of center cracked plate under biaxial loading. In this present study a perfect limit load can be evaluated for a various biaxial stress ratio. Lim et al. [8] presented fracture analysis of cracked orthotropic plate under biaxial loading. In the present study it is confirmed that only singular term is not sufficient to predict the crack extension. Goldstein and Shifrin [9] solved first mode crack deviation of orthotropic plane under biaxial loading where, a crack is considered as a thin elliptical hole. In this present work condition for crack stability and crack deviation is obtained and the present results are compared with the results of the conventional model of the crack, where crack is considered as ideal cut. Khatri and Lal [10] presented the stochastic fracture analysis of isotropic plate with hole and emanating crack under biaxial loadings by XFEM. In this present study different stochastic methods are used to see the seniority of different random variables in fracture analysis.

In the present paper MMSIF of center cracked FGM plate under bi-axial tensile load is evaluated for different fracture parameter like crack length and crack angle. XFEM based fracture analysis is utilized in the present study and the result obtained from present approach is also validated with the results from literature.

2. Problem formulation

In conventional FEM it is necessary to update mesh while solving fracture problem. This difficulty can be avoided in XFEM where updation of mesh is not required . The displacement vector in XFEM is written as.

$$u^h(x)=\sum_{i=1}^{n}N_i(x)\overline{u_i}+\sum_{i=1}^{n_c}N_i(x)H(x)a_i+\sum_{i=1}^{n_{t1}}N_i(x)\sum_{\alpha=1}^{4}\Phi_\alpha^1(x)b_i^{\alpha 1}+\sum_{i=1}^{n_{t2}}N_i(x)\sum_{\alpha=1}^{4}\Phi_\alpha^2(x)b_i^{\alpha 2} \tag{1}$$

Where $\overline{u_i}$, a_i $b_i^{\alpha 1}$ and $b_i^{\alpha 2}$ are the conventional degrees of freedom (dof), dof for enrichment function, crack tip and crack face asymptotic functions. A body with crack of area Ω and its outer boundary Γ is shown in Figure 1. The body is having volume / body loads b and the surface traction at the boundary Γ_t. The Boundary surface is $\Gamma=\Gamma_u+\Gamma_t+\Gamma_c$, where u , t and c are displacement, traction and discontinuity.

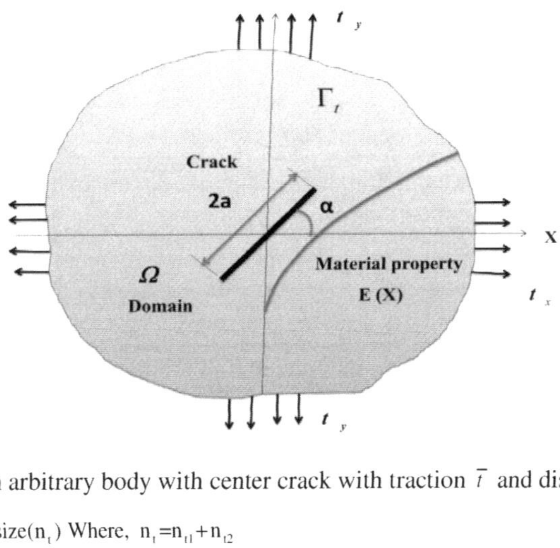

Figure 1: An arbitrary body with center crack with traction $\bar{t}$ and displacement $\bar{u}$

$$\text{dofs}=\text{size(n)}+\text{size(n}_c)+\text{size(n}_t)\ \text{Where,}\ n_t=n_{t1}+n_{t2} \tag{2}$$

Where, n , n_c and n_t represent the nodes for FEM mesh , crack face and crack tip respectively. $H(x)$ is the Heaviside function for the enrichment and the crack tip asymptotic functions are denoted by Φ_α^1 and Φ_α^2.

$$\Phi_{\alpha}^{t1} = \sqrt{r}\sin\left(\frac{\theta}{2}\right), \Phi_{\alpha}^{t2} = \sqrt{r}\cos\left(\frac{\theta}{2}\right) \tag{3}$$

$$\Phi_{\alpha}^{t3} = \sqrt{r}\sin\theta\cos\left(\frac{\theta}{2}\right), \Phi_{\alpha}^{t4} = \sqrt{r}\sin\theta\cos\left(\frac{\theta}{2}\right) \tag{4}$$

The J-integral in the form of SIF is represented as,

$$J = \frac{K_I^2 + K_{II}^2}{E_{eff}} \; Where, E_{eff} = \begin{cases} E & \text{for plane stress} \\ \frac{E}{1-\nu^2} & \text{for plane strain} \end{cases} \tag{5}$$

The J integral [10] for the cracked body is given in equation (6). The two states of J is represented by equation (7).

$$J = \int_{\Gamma}\left(W\delta_{1j} - \sigma_{ij}\frac{\partial u_i}{\partial x_1}\right)n_j d\Gamma \tag{6}$$

$$J^{(1+2)} = \int_{\Gamma}\left[\frac{1}{2}\left(\sigma_{ij}^{(1)} + \sigma_{ij}^{(2)}\right)\left(\varepsilon_{ij}^{(1)} + \varepsilon_{ij}^{(2)}\right)\delta_{1j} - \left(\sigma_{ij}^{(1)} + \sigma_{ij}^{(2)}\right)\frac{\partial(u_i^{(1)} + u_i^{(2)})}{\partial x_i}\right]n_j d\Gamma \tag{7}$$

On further solving, we get,

$$J^{(1+2)} = J^{(1)} + J^{(2)} + \frac{2}{E_{eff}}(K_I^{(1)}K_I^{(2)} + K_{II}^{(1)}K_{II}^{(2)}) \tag{8}$$

Now, comparing the Equations. (5) and (8) we get the interaction integral ,

$$I^{(1,2)} = \frac{2}{E_{eff}}(K_I^{(1)}K_I^{(2)} + K_{II}^{(1)}K_{II}^{(2)}) \tag{9}$$

The SIF in XFEM for two states are evaluated by putting $K_I^{(2)} = 1$ and $K_{II}^{(2)} = 0$ and $K_I^{(2)} = 0$ and $K_{II}^{(2)} = 1$ in equation (9), we get,

$$K_I^{(1)} = \frac{M^{(1,Mode\ 1)}E_{eff}}{2} \qquad K_{II}^{(1)} = \frac{M^{(1,Mode\ II)}E_{eff}}{2} \tag{10}$$

The following normalized mean MMSIF for first mode SIF and second mode SIF under tensile loading are represented as

$$K_I = \bar{K}_I / \sigma\sqrt{\pi a} \text{ and } K_{11} = \bar{K}_{II} / \sigma\sqrt{\pi a} \tag{11}$$

3. Result and Discussion

In present study a FGM plate with dimension (W = L = 1) under biaxial tensile loading is studied. In Figure 2 center crack FGM plate under Bi-axial loading, X-FEM meshing and enrichment for crack tip and face are shown. The material properties for FGM plate are $E_1 = 10^4$, $E_2 = 10^3$, $\nu = 0.3$. Equation 12 represents the change of material properties along width.

$$E_1(x) = E_1 e^{\beta x}, \quad E_2(x) = E_2 e^{\beta x} \quad G_{12} = \frac{0.5\,E}{1+\upsilon} \tag{12}$$

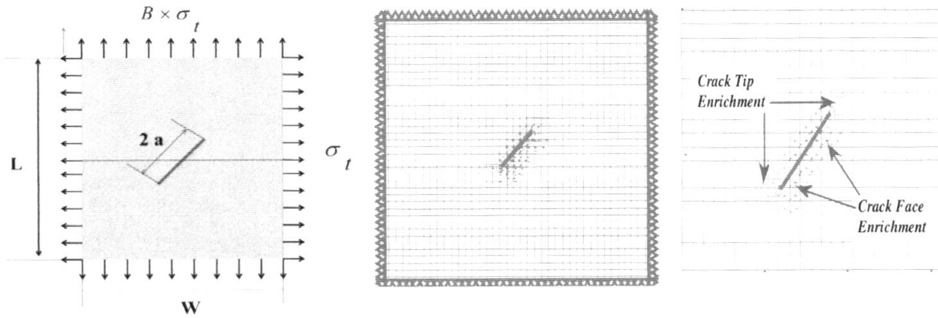

(a) (b) (c)

Figure 2: Plate under biaxial tensile stress (a) geometry and loading condition (b) X-FEM meshing of the plate with centre inclined crack (c) enrichment for crack tip and crack face by X-FEM.

In Figure 3 validation study of FGM plate with center crack of crack length (a=0.1) under uniform uniaxial loading is presented, where MMSIF K_I is calculated for gradation factor (β). It is observed that result of present model is very close to the result from the literature.

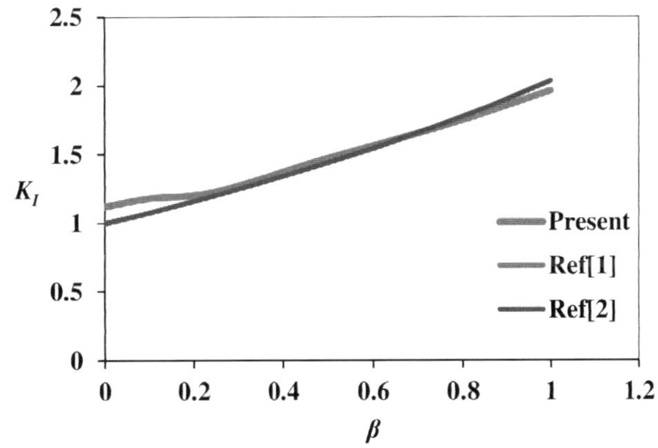

Figure 3: MMSIF K_I for center cracked FGM plate subjected to uniaxial tensile loading

In Table 1 MMSIF K_I and K_{II} of the center crack isotropic plate under uniform uni-axial loading is calculated and this result is compared with the result of FGM plate of same dimension and loading. It is observed that MMSIF K_I and K_{II} in FGM plate is less as compared to Isotropic plate. So, from this study it is concluded that FGM provides better fracture resistance than Isotropic material. The maximum reduction of K_I and K_{II} is observed as 4.41 % and 15.38 % respectively.

Table 1: MMSIF in Isotropic plate and FGM (β=0.1) plate under uni-axial loading.

a	MMSIF	Isotropic Plate	FGM Plate
0.1	K_I	1.202	1.188
	K_{II}	0.003	0.002
0.15	K_I	1.284	1.283

	K_{II}	0.009	0.008
0.2	K_I	1.360	1.340
	K_{II}	0.022	0.019
0.25	K_I	1.495	1.429
	K_{II}	0.039	0.033

In Figure 4 MMSIF K_I of center crack FGM plate for a = 0.1 and α = 0 with respect to β under uni-axial and bi-axial loadings is presented. In this study it is observed that the value of MMSIF K_I is get reduced in bi-axial (B=1) loading as compared to uni-axial loading of same magnitude. This is due to loading in lateral direction which reduces the effect of loading in Y direction. The maximum reduction in MMSIF K_I is 8.21 % for β =0, where as the minimum reduction is 2.24 % for β=1. This study is further extended to the effect of bi-axial (B=2) loading and it is observed that for center cracked FGM plate the value of MMSIF K_I is get almost double as compared to the bi-axial (B=1) loading.

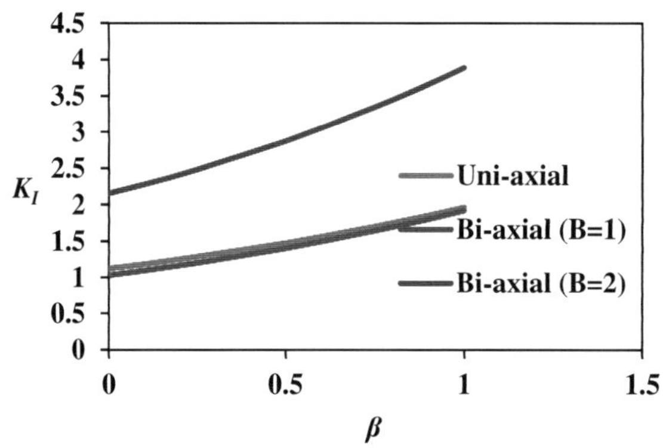

Figure 4: MMSIF K_I for center cracked FGM plate subjected to uniaxial tensile loading for different values of β.

In Figure 5 MMSIF K_I of center cracked FGM plate for β=0.25 and α=0 with respect to a under uni-axial and bi-axial loadings is presented. In this study it is observed that the value of MMSIF K_I is get reduced in bi-axial (B=1) loading as compared to uni-axial loading of same magnitude. This is due to loading in lateral direction which reduces the effect of loading in Y direction. The maximum reduction in MMSIF K_I is 84.76 % for a = 0.3, where as the minimum reduction is 6.77 % for a=0.1. This study is further extended to the effect of bi-axial (B=2) loading and it is observed that for center crack FGM plate the value of MMSIF K_I is get increased with maximum 7.13 times for a= 0.3 and minimum 2.07 times for a = 0.1.

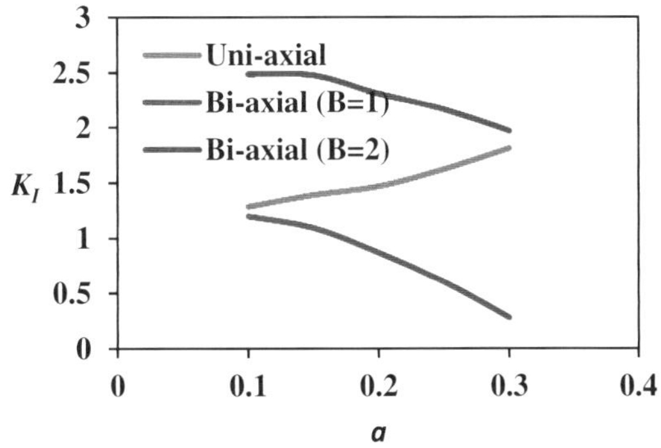

Figure 5: MMSIF K_I for center cracked FGM plate under uni-axial and Bi- axial loadings for different values of a

In Figure 6 MMSIF K_I and K_{II} for center crack FGM plate where $\beta = 0.25$, $a = 0.2$ are calculated for different values of crack angle (α) under uni-axial and bi-axial (B=1 and B=2). In this study it is observed that for uni-axial loading as crack angle increases from 0^0 to 50^0 MMSIF K_I decreases by 59.19 % whereas K_{II} increases by 31.85 times . This is due to shearing component of uni-axial tensile load. In center cracked FGM plate under bi-axial stress it is observed that as crack angle increases from 0^0 to 50^0 MMSIF K_I increases where as maximum value of K_{II} is observed at 40^0 crack angle. It is also observed that as crack angle increases from 0^0 to 50^0 the MMSIF K_I is more and K_{II} is less in bi-axial (B=2) loading as compared to bi-axial (B=1) for all values of crack angle.

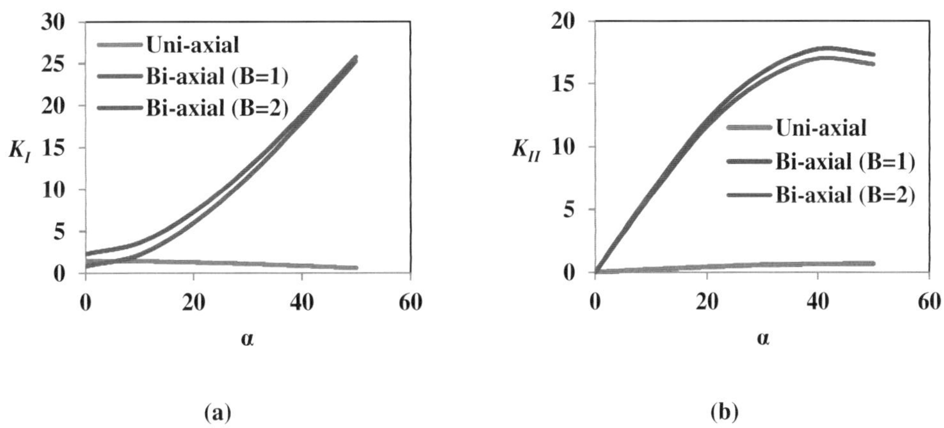

(a) (b)

Figure 6: MMSIF K_I and K_{II} for center cracked FGM plate under uni-axial and Bi- axial loadings for different values of α

4. Conclusions

In this present work MMSIF of center cracked FGM plate under bi-axial loading is presented. The mathematical formulation is done by XFEM using J –integral in MATLAB. The conclusions of this present work are summarized as

- MMSIF K_I and K_{II} in FGM plate is less as compared to Isotropic plate. This is due to better fracture resistance in FGM material than Isotropic material. The maximum reduction of K_I and K_{II} is observed as 4.41 % and 15.38 % respectively when crack length varies from 0.1 to 0.25.

- It is observed that the value of MMSIF K_I is get reduced in Bi-axial (B=1) loading as compared to uni-axial loading of same magnitude. This is due to loading in lateral direction which reduces the effect of loading in Y direction. It is also observed that MMSIF K_I is get almost double for bi-axial (B=1) loading as compared to the bi-axial (B=1) loading when gradation factor varies from 0.1 to 1.
- It is observed that the value of MMSIF K_I is get reduced in Bi-axial (B=1) loading as compared to uni-axial loading of same magnitude as crack length (a) varies from 0.1 to 0.3. This is due to loading in lateral direction which reduces the effect of loading in Y direction.
- In this study it is observed that for uni-axial loading as crack angle increases from 0^0 to 50^0 MMSIF K_I decreases by 59.19 % whereas K_{II} increases by 31.85 times .

References

[1] Kim J H, Paulino G H 2003 *International journal of solids and structures,* **40,** 3967-40001
[2] Bayesteh H and Mohammadi S 2013 *Journal of Composites Part B,* **44,** 8-25
[3] Lal A. and Palekar S P 2015 *Int J Mech Mater Des.,* **13-2,** 195-228
[4] Mohammadi S 2012 *John Wiley & Sons Ltd. Publication*
[5] Ebrahimi S H, Asadpoure A and Mohammadi S 2007 *Media Int J Civil Eng,* **6-3,** 198-207
[6] Lal A, Mulani S B and Kapania R K 2017 *International journal of applied mechanics,* **9-4,** 1750061(1-33)
[7] Meek C and Ainsworth 2015 *Eng. Fract. Mech.*, **147,** 306–317
[8] Lim W K, Choi S Y and Shankar B V 2001 *Eng. Fract. Mech.*, **68,** 403-416
[9] Goldstein R V and Shifrin E I 2012 Int. J. Eng. Sci., **61,** 36–47
[10] Khatri K and Lal A 2018 *Theoretical and applied fracture mechanics,* **96,** 1-22

IOP Conf. Series: Materials Science and Engineering 1004 (2020) 012022 doi:10.1088/1757-899X/1004/1/012022

Experimental investigation of cylindrical roller bearing for inner race defect under varying load

Sachin P. Patel [1], **S. H. Upadhyay** [2*]

[1]Research Scholar, Department of Mechanical & Industrial Engineering, IIT Roorkee, Roorkee-247667, Uttarakhand, India.
[2]Associate Professor, Department of Mechanical & Industrial Engineering, IIT Roorkee, Roorkee-247667, Uttarakhand, India.
[*]Email: sanjay.upadhyay@me.iitr.ac.in

Abstract. This paper presents an experimental investigation of inner race defect on cylindrical roller bearing under various speeds and radial load conditions. The several statistical features like RMS, crest factor, skewness, kurtosis, and frequency domain techniques like FFT, envelope spectrum have been extracted and compared it for defective and healthy bearing at various speeds and radial load to understand the performance of bearing. The experimental results indicate that increment in rotational speed and variation in radial load has a remarkable influence on various features of the signals.

1. Introduction

Roller bearings have been extensively used in various industrial machineries since the late nineteenth century. It requires constant monitoring to avoid unexpected failure. Any misalignment or faults produced in any part of bearing can cause vibration, which magnifies with the hike in load and speeds, and results in complete failure of bearing or machinery.

Some experimental research was carried out with rolling element bearings. The experimental investigation has performed by [1] for ball bearing defect with and without unbalancing mass. The effect of surface roughness damage under radial load at different positions of the outer ring of ball bearing has been studied by [2]. [3] has experimented for the fault on inner ring, outer ring, and ball considering varying speed and radial load. The experimental examination for ball bearing has been performed for dynamic motions of cage under different speeds and loads by [4]. The influence of individual and combined localized defects with variations in defect size, rotational speed, and radial load has been studied by [5]. Faulty ball bearing running at various speeds and many radial loads with several fault sizes has been analyzed by [6].

The RMS value can be consider for bearing diagnosis as an indicator of an average of overall amplitude of vibration [3]. Its value is always less than the peak value. Kurtosis is a fourth-order moment, which describes the heaviness of the tails and the peakedness of the distribution. High kurtosis value shows that high impulsive components are present in the signal [7]. Skewness measures the symmetricity of the dataset. '0' skewness indicates perfectly symmetrical data points. Skewness from -0.5 to 0.5 shows the fairly symmetrical region. Skewness from -0.5 to -1, or from 0.5 to 1 point out the moderately skewed region. Skewness below -1 or above 1 has a highly skewed region. Crest factor measures the impulsive nature of the signal. It is found by dividing the peak of the waveform by RMS. Fourier analysis transforms a signal from time-domain to frequency-domain or vice versa. A fast Fourier transform (FFT) shows frequencies present in the vibration signal. Envelope spectrum is an effective technique to detect defect frequency of bearing. However, it is not suitable for healthy bearing.

2. Experimental / Computational details

In an experimental setup, shaft is supported by two bearings. One of them is a support bearing and another is a test bearing. A 3-Φ, 3000 rpm induction motor was connected to the shaft with a flexible coupling. A 3-Φ variable frequency drive was linked with motor to control its speed. A laser tachometer was adjusted to observe the rotational speed of the shaft. The entire setup was placed on an aluminium plate having six vibration absorber pads. The artificial fault was formed on the inner race of bearing by EDM.

The bearing vibrations were acquired by piezoelectric accelerometers 352C03 with sensitivities of 10.07 mV/g; which was fitted in the vertical direction on test bearing housing. A DAQ card NI 9234 was utilized to acquire the vibration signal. Data collection was conducted through NI LabVIEW program. The sampling rate was kept to 25.6 kHz. The data was collected for a duration of 1 sec. Collected data were then loaded into MATLAB for analysis.

During the experiment, the outer race was kept motionless in bearing housing, and the inner race rotates along with the shaft. Cylindrical roller bearing NJ 305E with inner race defect and without defect were used for the experiment. Details of the bearing are registered in Table 1.

Table 1: Roller bearing specifications

Properties	Value	Properties	Value
Number of rollers (n_r)	11	Inner race radius (I_r)	17 mm
Diameter of roller (d_r)	10 mm	Width of defect (W)	13 mm
Length of roller (l)	11 mm	Depth of defect (D)	0.5 mm
Pitch diameter (P_d)	43.65 mm	Outer race radius (I_o)	27 mm

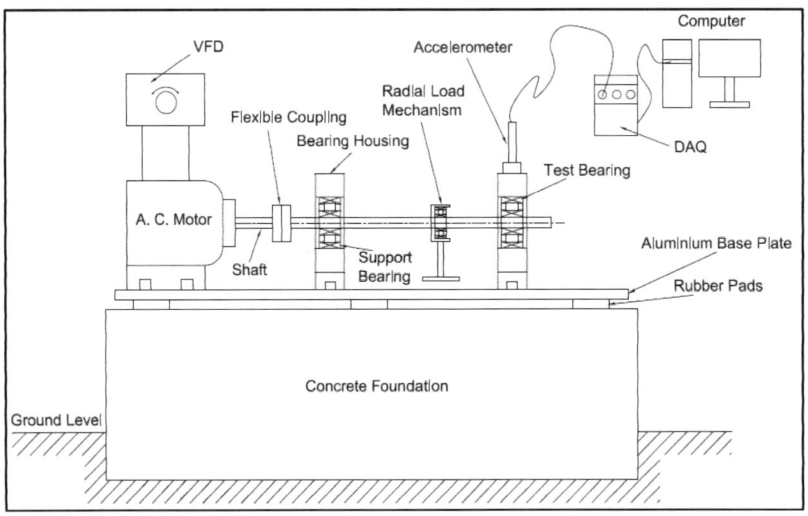

Figure 1: Schematic diagram of experimental setup.

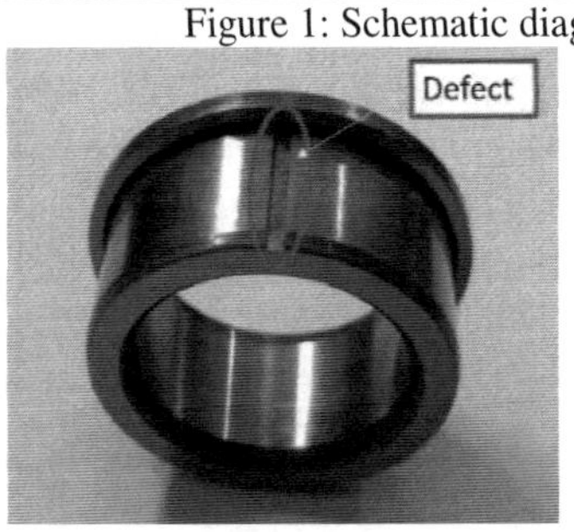

(a) Defect (b) Defect creation through EDM

Figure 2: Inner race defect

The experimental tests were carried out for inner race fault of size 0.5 mm defect length and for healthy bearing. It is also tested at various speeds in the range of 300 to 3000 rpm with a speed increment of 300 rpm. The bearings were tested for 0 and 3 kg radial load.

Table 2: Various frequencies in rad/sec

Frequency	Formula
Shaft frequency	$\varphi_s = \frac{2.\pi.n}{60}$
Cage frequency	$\varphi_c = \frac{\varphi_s}{2} \times \left(1 - \left(\frac{d_r}{P_d}\right)\right)$
Varying compliance (VC) frequency	$VC = n_r \times \varphi_c$
Inner race defect (IRD) frequency	$\varphi_i = n_r \times (\varphi_s - \varphi_c)$

Table 3 Time-domain features [8]

$RMS = \sqrt{\frac{1}{P}\sum_{i=1}^{P} x_i^2}$	$skewness = \frac{\frac{1}{P}\sum_{i=1}^{P}(x_i - \bar{x})^3}{(std.dev.)^3}$
$kurtosis = \frac{\frac{1}{P}\sum_{i=1}^{P}(x_i - \bar{x})^4}{(std.dev.)^4}$	$crest\ factor = \frac{max\lvert x_i\rvert}{RMS}$

3. Result and Discussion

The vibration signals were acquired at various speeds for with and without load along with faulty and healthy bearing. Out of numerous vibration signals, the signals acquired for defective and healthy bearing at 900 rpm are displayed in Figure 3, and Figure 4. The periodic impacts occur at defect frequency of inner race, which can be estimated from Table 2. The shaft frequency and varying compliance frequency are also present in the system.

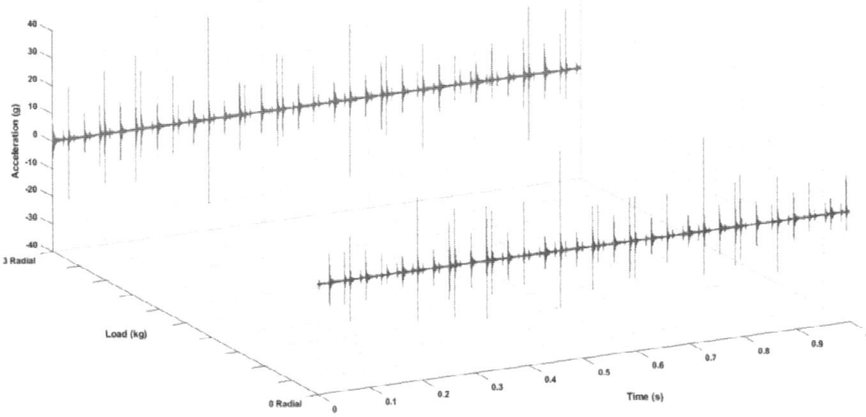

Figure 3: Time signal of IRD bearing

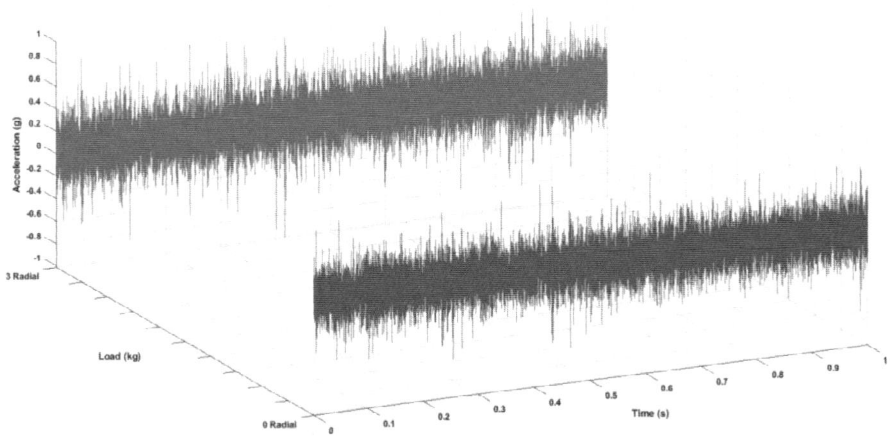

Figure 4: Time signal of Healthy bearing

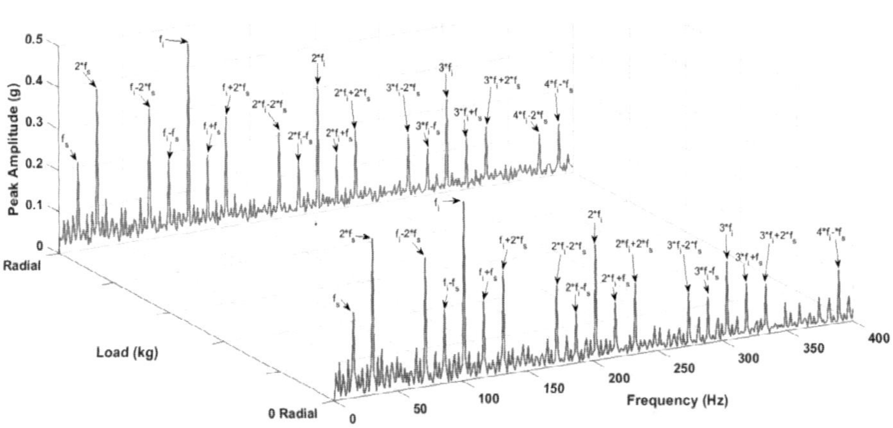

Figure 5: Envelope spectrum for IRD

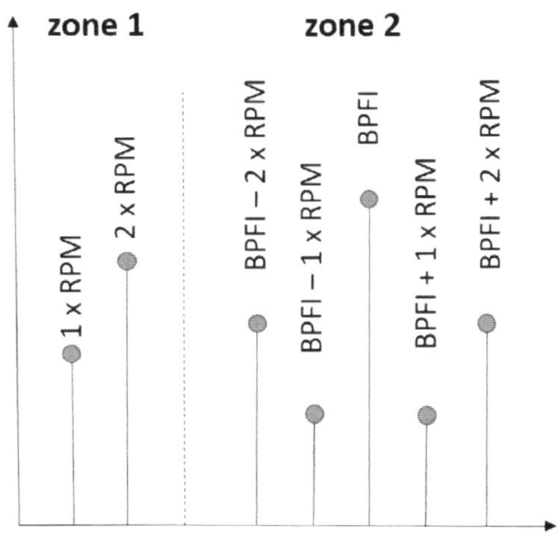

Figure 6: Envelope spectrum pattern for IRD

Figure 5 shows the envelope spectrum for bearing running at 900 rpm with inner race defect. Due to manufacturing limitations, there is always some clearance between the side flange of races and roller side face. Due to that clearance, inner race fluctuates axially with respect to outer race by some amount. This phenomenon results in the formation of periodic shaft frequency as in zone 1 and sideband frequencies around defect frequency of BPFI ± 1 x RPM and BPFI ± 2 x RPM as in zone 2, which is clarified in Figure 6.

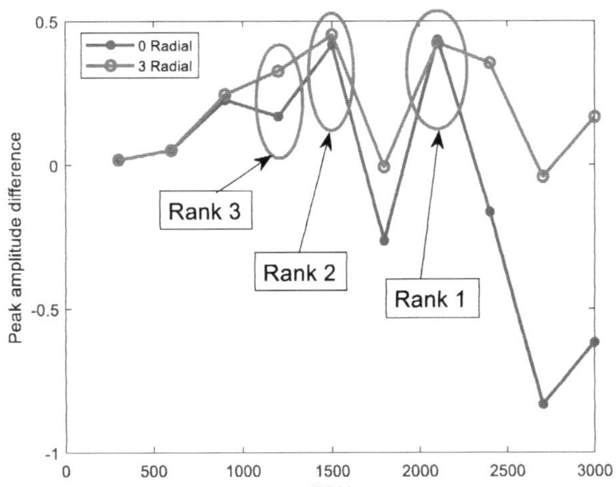

Figure 7: Envelope spectrum amplitude difference between defect frequency and shaft frequency

Figure 7 shows the difference between the amplitude of defect frequency and shaft frequency at different speeds. The higher difference indicates that defect frequency has a major role, and the lower difference indicates shaft frequency has a major role in vibration signal. So, based on this, rank has been decided for the selection of optimum speed at which defect frequency is predominant. So, as per Figure 7, 2100rpm, 1500rpm & 1200rpm has been assigned Rank 1, Rank 2, and Rank 3, respectively.

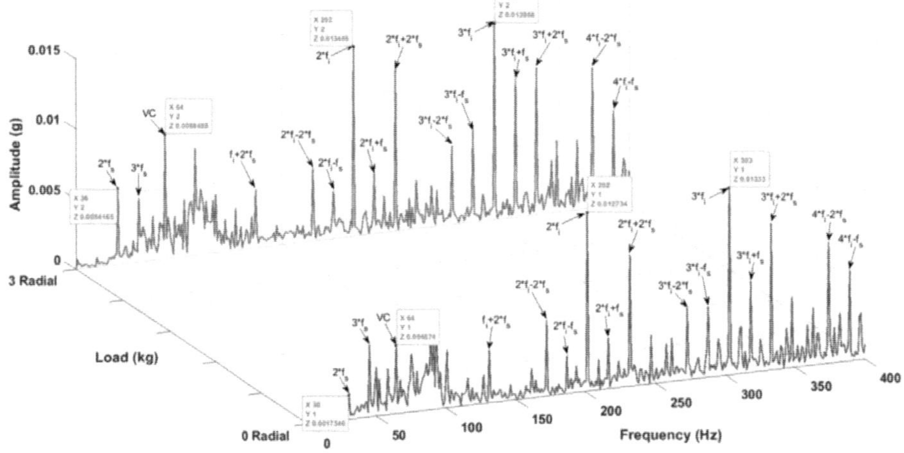

(a)

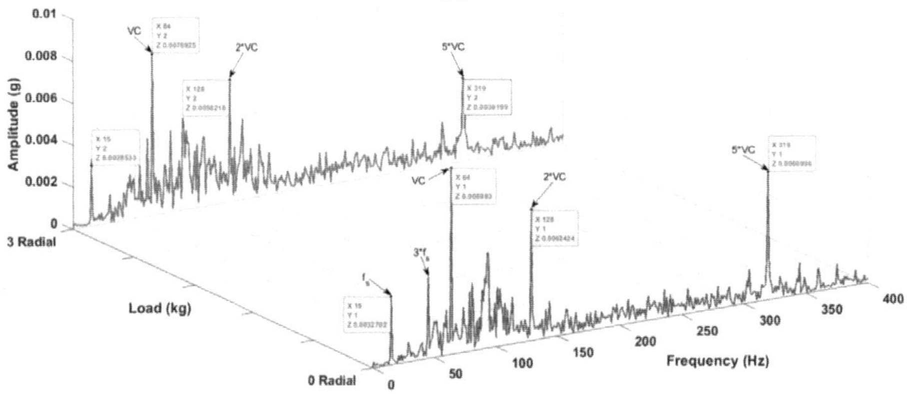

(b)

Figure 8: FFT plot for (a) IRD, (b) Healthy at 900 rpm

Figure 8-a displays shaft frequency and defect frequency along with their harmonics for defective bearing. Whereas, Figure 8-b shows shaft frequency and Varying Compliance (VC) frequency along with their harmonics for healthy bearing. By carefully observing Figure 8, it seems that the defective bearing has a higher frequency amplitude compared to the healthy bearing. In Figure 8-b, with no load condition shaft frequencies 'f_s' and '$3f_s$' can be observed; however, with the application of radial load, only 'f_s' is visible, which indicates that radial load reduces the effect of shaft frequency for healthy bearing.

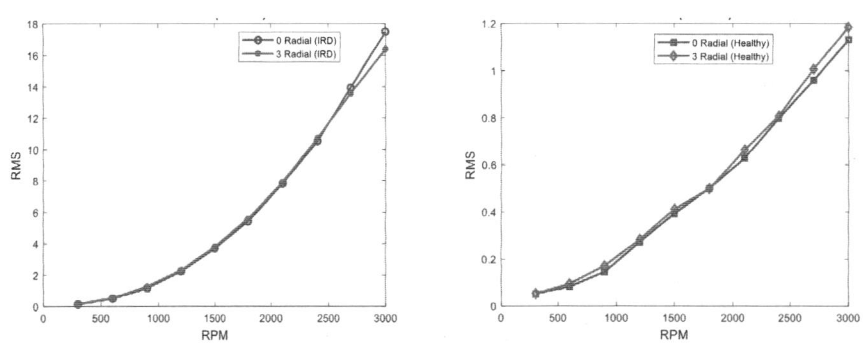

(a) (b)
Figure 9: RMS for (a) IRD, (b) Healthy

From Figure 9 it confirms that the defective bearing has higher RMS value than healthy bearing at all speeds. With increase in speed RMS value increase for healthy and faulty bearing. At a higher speed, the no-load condition has higher RMS value compared to with load condition for defective bearing, while for healthy bearing, it is opponent.

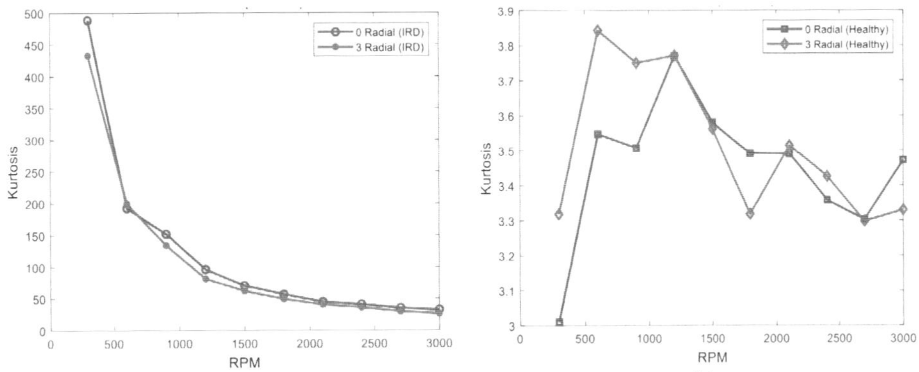

(a) (b)
Figure 10: Kurtosis for (a) IRD, (b) Healthy

The defective bearing has a higher kurtosis value than healthy bearing, which is witnessed from Figure 10. For healthy bearing, kurtosis is nearby 3. As shown in Figure 10, kurtosis is high at low speed and reduces continuously with increment in speed for the specified defect size, while for non-defective bearing, it increases initially and then decreases. Under the situation of 0 radial load, higher kurtosis value is observed compared to 3 kg radial load for defective bearing at all speeds.

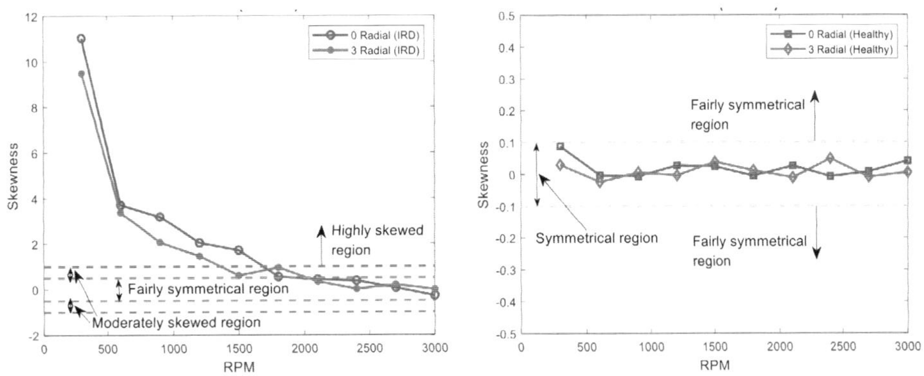

(a) (b)
Figure 11: Skewness for (a) IRD, (b) Healthy

As illustrated in Figure 11, the defective bearing has higher skewness value than the healthy bearing. With the increment in speed, skewness value decreases and approaches towards symmetricity for faulty bearing, which is described in Figure 11-a. For defective bearing up to 1500 rpm data set falls into the highly skewed region, from 1800 to 2100 rpm moderately skewed region and from 2400 to 3000 rpm fairly symmetric region. In highly skewed region, bearing without load has higher skewness value compared to with load condition. When the dataset falls into a moderately or fairly symmetrical zone, then skewness value fluctuates between no load and with load condition. Which confirms that data points in a highly skewed region will have more effect of loading than moderately or fairly symmetrical region.

Figure 11-b indicates that for healthy bearing skewness for all speeds are close to 0, which confirms that for healthy bearing perfectly symmetrical data set will get irrespective of speed.

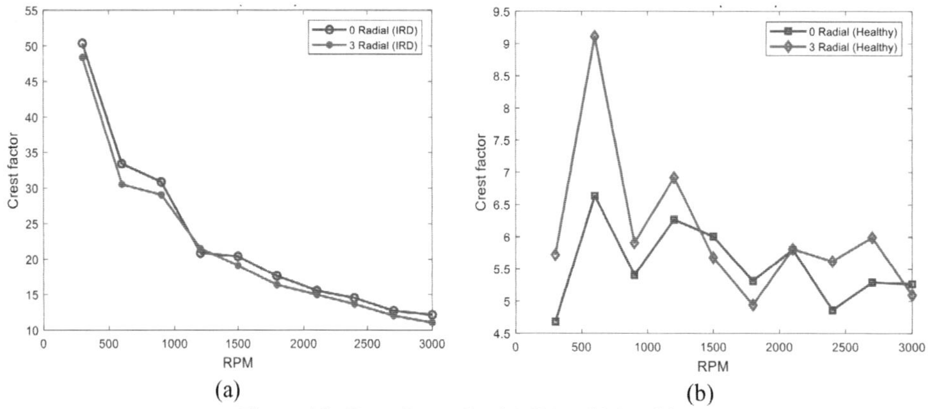

Figure 12: Crest factor for (a) IRD, (b) Healthy

Figure 12 shows that defective bearing has a higher crest factor compared to the healthy bearing. For healthy bearing, the crest factor is around 6. From Figure 12-a it can say that for defective bearing crest factor reduces with an increase in speed, however, as per Figure 12-b for healthy bearing, it fluctuates. The no-load condition has a higher crest factor compared to with load condition for defective bearing at all speeds, while for healthy bearing crest factor varies among no-load and with load condition.

4. Conclusion

- The selection of speed based on the predominance of defect frequency can be achieved from a peak amplitude difference between defect frequency and shaft frequency.
- The loaded bearing has lower frequency amplitude compared to no-load condition for healthy bearing, while for defective bearing, it is contrariety.
- Radial load reduces the effect of shaft frequency for healthy bearing.
- Results confirm that radial load has antagonist behavior between defective and healthy bearing for RMS value at a higher speed.
- Radial load helps to reduce kurtosis value for defective bearing.
- For healthy bearing, skewness falls in the symmetrical region irrespective of speed. For defective bearing data points in a highly skewed region will have more effect of loading than other regions.
- With the increment of speed, the signal approaches towards symmetricity from the highly skewed region.
- Radial load acting on bearing results into a reduction in crest factor for defective bearing. For healthy bearing at low speeds, the radial load has a major effect on the crest factor.
- For healthy bearing kurtosis, skewness, and crest factors are around 3, 0, and 6, respectively; however, these features increase drastically with damage.
- Among all time-domain features described here, skewness has a significant influence on various loading conditions for the specified defect size.

References

[1]Nagane S S, Kulkarni S S and Gandhare B S 2014 Experimental investigation of rolling element bearing with unbalanced rotor *International Journal of Engineering Research & Technology,* **03,** (05), 1695-1701

[2]Kulkarni S and Wadkar S B 2016 Experimental investigation for distributed defects in ball bearing using vibration signature analysis *Procedia Engineering,* **144,** 781-789

[3]Saruhan H, Saridemir S, Cicek A and Uygur I 2014 Vibration analysis of rolling element bearings

defects *Journal of Applied Research and Technology,* **12** (03), 384-395
[4]Wen B, Ren H, Zhang H and Han Q 2017 Experimental investigation of cage motions in an angular contact ball bearing *Proceedings of the Institution of Mechanical Engineers, Part J: Journal of Engineering Tribology,* **231** (08), 1041-55
[5]Patel J P and Upadhyay S H 2015 Experimental studies of vibration due to combined defect of roller bearings using response surface methodology *Journal of Vibration Analysis, Measurement, and Control,* **03** (02), 74-92
[6]Shaha R D, Kulkarni S S and Gandhare B S 2014 Experimental investigation of effect of local defect on deep groove ball bearing under pure radial load *International Journal of Mechanical Engineering And Information Technology,* **02** (06), 308-316
[7]Liu S, Hou S, He K and Yang W 2018 L-Kurtosis and its application for fault detection of rolling element bearings *Measurement,* **116,** 523-532
[8]Rai A and Upadhyay S H 2018 Intelligent bearing performance degradation assessment and remaining useful life prediction based on self-organising map and support vector regression *Proceedings of the Institution of Mechanical Engineers, Part C: Journal of Mechanical Engineering Science,* **232** (06), 1118-32

IOP Conf. Series: Materials Science and Engineering 1004 (2020) 012023 doi:10.1088/1757-899X/1004/1/012023

Slurry erosion behaviour of Ni-Hard under various impact angle and speed

Mohit Makwana[1], B. M. Sutaria[2]

[1, 2] Mechanical Engineering Department, Sardar Vallabhbhai National Institute of Technology, Surat, India

Abstract. To Study wear analysis of slurry pump with an experimental condition is an economically costly and lengthy process. Therefore, wear study can be done on various laboratory tests such as Slurry pot tester, Coriolis wear tester, Whirling arm wear test device, particle impact jet test. To analyze of erosion wear rate of slurry pump material in different operating conditions is performed on slurry pot tester. In the latest pump the Ni-Hard is widely used material in the slurry transportation. The one of the major application of slurry pump made from Ni-Hard is in Thermal power plant for ash slurry transportation. In this paper an experimental work has been carried out to find the effect of impact angle and particle impact speed at 20% w/w slurry concentration of sand on the wear rate of the Ni-Hard material. From the experimental results it is found that as particle velocity increases erosion wear rate of the material is increases. It was also found that the erosion of the material increases with impact angle till 30° then the erosion of material decreases gradually.

1. Introduction

Slurry transport is essential in many industries such as mining, thermal power plant, chemical, petroleum. Solid particles to be transported in the slurry form for small to medium distances. Various industries have a large number of applications for slurry pumping. The slurry pump is the most important component of the slurry transport system. The life of the slurry pump varies with the range of a few weeks to a few years depending upon particle size distribution, slurry concentration, abracivity of the particle, specific density of particle etc. The main problem regarding slurry transport is low life expectancy and degradation of pump performance due to erosion. Hence increasing the downtime and reducing productivity. Erosion wear is important very important part for design and selection of the slurry pump Satish R More et al. [1]. S. Bree et al. [2] observed that with increase in particle kinetic energy erosion wear also increases.

To calculate and study the erosion occurred in the slurry handling materials a high speed slurry pot tester was developed by Ojala et al.[3]. The test sample is attached to whirling arm of the driven by the rotating shaft. The maximum particle size of 10mm can be used in this setup. Experiment on high speed erosion was done by Singh et al.[4] where they studied the erosion characteristic of 75 μm of particle size fly ash on the cast iron impeller. Desale et al.[6] developed a slurry pot tester for determining effect on erosion by various characteristic of slurry. Where they determined with the increase in particle concentration erosion wear also increases. The slurry pot tester developed by Desale et al. [5] can be used for various particles sizes, concentration, and impact angle. Desale et al.[6] studied the effect the slurry concentration on the erosion wear where they found that erosion increases as concentration increase. Ni-Hard is mostly used in slurry pumping applications, cylinder

IOP Conf. Series: Materials Science and Engineering 1004 (2020) 012023 doi:10.1088/1757-899X/1004/1/012023

liners, pistons, bearings, glands. Here in this study effect of the impact angle and impact speed on the erosion of Ni-Hard material at constant concentration and particle sized is observed.

2. Experimental Setup

A pot tester of 7 liters capacity developed by More et al.[1] is used in the present study. Figure 1 shows the schematic diagram of the slurry pot tester is shown. This Slurry pot has a capacity of containing 7 liters of water for slurry erosion test. To mitigate the vortex motion generated by the rotation of arm of the slurry pot and bottom propeller four baffles are provided. One drain hole in the bottom of the slurry pot and one hole on the acrylic lid to fill the slurry so no air remain in the slurry pot to interfere in the erosion. The simple design is useful for conducting the experiments as factors involving the slurry erosion can be set as per the requirement.

2.1 Preparation of Test Fixtures and Samples

The erosion wear test in the slurry pot testing machine requires the preparation of test specimens with proper dimension to fix in to the test fixtures. The preparation and mounting of the test sample in to the test fixture and attachment of the test fixture to motor through upper shaft is shown in Figure 1.

2.1.1 Test Fixture

Special test fixtures were fabricated to mount flat wear specimens. The erosion wear of the fixture is negligible which can impact on the performance or the accuracy of the test for the amount of time test is running. That is why same fixture can be used for the several test runs. A dimension of the sample ise 30 mm x 5 mm x 2 mm, rounded at two ends was provided at the test fixture to fix a wear specimen inside it. Experiments on different angles to study the effect of impact angle is provided at every 15° from 0° to 90°. The slotted angular plate is welded to the holding arm being fitted with the brass sleeve. The brass sleeve is fixed to the upper shaft in such a way that the centre of the wear specimen could be placed equidistance from the top cover and the propeller (approximately 65 mm from each one). The test fixture, one on each holding arm, is hanged opposite by a 4 mm screw. Two fixtures were fitted at 180° apart to balance the dynamic forces, if any, with minimum wake interference-effect. It is expected that this arrangement will help in orienting the wear specimens at the angle of the tilt of the fixture for impact with the suspended solid particles. The test fixtures are rotated at a 71 mm radius in a direction opposite to the propeller rotation to minimize the rotational motion effect. Since the volume swept by the wear specimens and the holding arms is very small it can be said the condition are ideal during the set period of time.

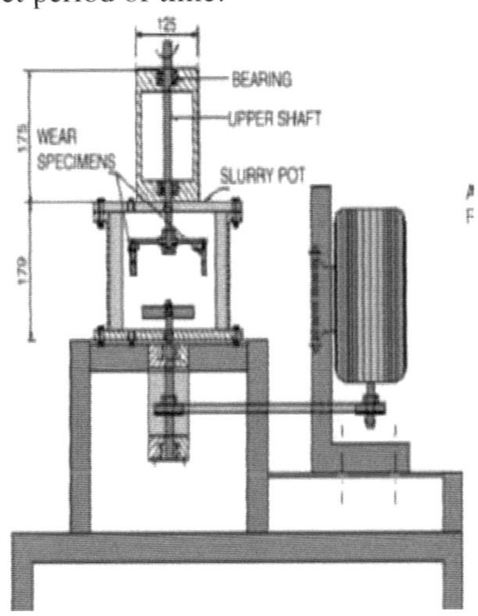

Figure 1: Schematic diagram of Slurry pot tester. [All dimensions are in mm] [Desale et al. 2005]

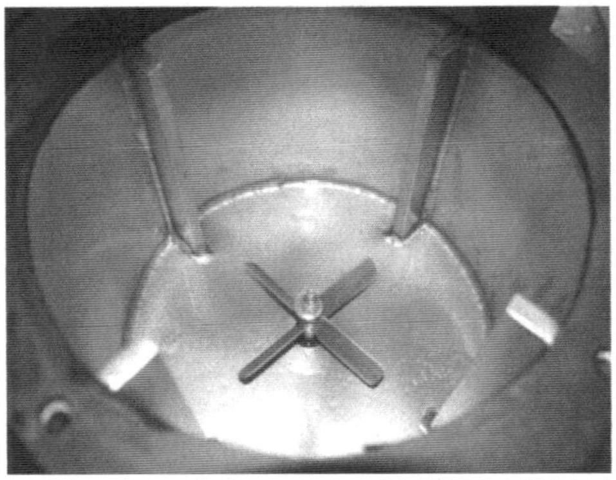

Figure 2: Photographic view of slurry pot with PBT 45° blade Desale et al [7]

2.1.2 Test Specimens

Ni-Hard (Chemical Name- Nickel Chromium Martensitic white Iron NiCr 30/550) being a cast iron used widely in ash slurry pump has been chosen as the wear test piece material for the present study. The size of the wear piece used was (30 mm x 5 mm), rounded at both ends with a thickness of 2 mm (surface are 144.64 mm2) to fit in the slot of the fixture. The wear pieces were polished using #110, #220, # 600 & # 1000 number emery papers and cloth before conducting the experiment to ensure identical initial conditions for each set of data. Then the test samples were cleaned with water and acetone for a clean surface.

2.2 Properties of Material Used

In the present investigation Indian Standard Sand is used as erodent material and Ni Hard (Type 1B) as target material.

2.2.1 Particle Size Distribution

It is well known that particle size of the particle in the slurry have considerable impact on the erosion wear of the slurry handling equipment. Therefore it is essential to know determine the particle size distribution of the erodent material for establishing the relation between different parameters. Particle sieve analysis is used to determine particle size distribution for coarser particles. This distribution has been obtained by the dry sieve analysis method. Here the sand of different sizes is passed through the number of sieves to categorise as per the size requirement. Here care is taken to keep the sand dry before mixing it with water. The sample retained on each sieve is collected, and its percentage is calculated following the standard procedure.

The particles of IS sand grade III are selected as erodent material for the investigation, and the physical properties are shown in Table 1. Collecting identical sand size is not possible because of fine size and amount of the material being used. The particles are, therefore, sieved using successive sieves sizes, and the mean sieve size designates the particles collected between two successive sieves. To collect quartz particles, Grade III of Indian standard (IS 650:1991) sand was used. From Grade III, the mean particle size of 655 μm was collected as the material retained between the two sieves of 710 μm and 600 μm sizes.

Table 1 Erodent Properties

Solid particle	Chemical formula	Colour	Sp. Gravity (Kg/m3)	Hardness VHN	Particle Shape
Quartz (IS Sand)	SiO_2	Whitish	2652	1100	Sub Angular

2.2.2 Properties of Target Material

The material was acquired from the Thermal Power Plant. The Ni-Hard has different types of material in various percentages depending on the composition of Iron, Nickel, Chromium, Silicon, Manganese etc. Here in this study Ni-Hard (Type 1B) is used and its elemental composition is given in Table 2. The hardness of the material was tested on Mututoyo Rockwell Hardness Testing Machine and found 62.5 HRC hardness of the material. Ni-Hard generally used for abrasion resistance and erosion resistance. This type of material used for the different types of slurry pump. Mainly it is used in ash handling plant of thermal power station to transport ash slurry. This material is used in the form as available i.e. machinable and no heat treatment is carried out. The table motioned below shows the typical range of the Ni-Hard (Type 1B) composition. Toughness and resistance to repeated shock increase as carbon content decreases.

Table 2 –Target material properties

Target Material	Elemental composition (wt. %)				
	C	Mn	P	S	Si
NI Hard (Type 1B)	2.8- 3.3	0.3-0.6	0.3	0.015 Max	1.5-2.2
	Cr	Ni	Mo	N	Cr
	8.0 -10.0	4.0-6.0	0.5 Max	0.10	8.0 – 10.0

2.3 Range of Parameters

The experiments have been carried out using sand slurry of IS sand of 655 μm mean particle size. The mixture was prepared by mixing the sand and tap water in the slurry pot. Slurry erosion of material depends on the many parameters such as particle size, concentration of slurry, impact velocity, impact angle, particle shape etc. Here in this study concentration has been taken 20% by weight and Mean particle size of 655 μm of sand has been taken as constant. Slurry erosion of with two impact velocity (8 and 10 m/s) with different impact angles (15, 30, 45, 60, 75, 90) has been taken as to study the effect of the impact angle and impact velocity.

2.4 Experimental Procedure and Data Analysis

The every test specimens for each experiment are polished with #110, #220, #600 and #1000 emery paper for mirror finish and cleaned with water and acetone respectively. Mass loss of the wear specimen is measured by an electronic balance having least count of 0.1 mg. Each experiment is done with 90 min time. The average mass loss of two wear specimens is used to evaluate the erosion rate according to the relationship proposed by Bree et al.

$$Ew = \frac{W_L}{\rho_S \times A_{SP} \times Cv \times V_{SP} \times T} \quad (1)$$

$$Cv = \frac{Cw}{\rho_S - (\rho_S - 1) \times Cw} \quad (2)$$

Where,

ρ_S - Mass density of the solid particle material, (kg/m3)

A_{SP} - Surface area of the wear specimen subjected to erosion, (m2)

C_v - Solid concentration by volume, fraction;

C_w - Solid concentration by weight, fraction;

E_w - Total erosion rate, (g/g)

T - Time over which mass loss has been measured, (sec)

V_SP - Peripheral velocity of wear specimen, (m/s)

W_L-Measured mass loss, (kg)

Preparation of the sample and slurry are key factors for the slurry pot testing. That is why test is done as a standard procedure by first mixing the slurry where predetermine amount of known mean particle size particles and water is mixed in the slurry pot. Sample is prepared by cleaning and polishing to the mirror finish without and external layer of oil or dirt. After closing the acrylic lid and first to ensure particles are uniformly distributed in the slurry bottom propeller is given and have to run at a desired speed after that main shaft should run to at a desired experimental requirement.

3. Results and Discussion

The impact angle of the particles is one of the critical parameters affecting the erosion rate of the target material. It gives an idea about the location of maximum erosion in slurry handling equipment. Hence initially, the erosion wear of the materials was evaluated at different impact angles in the pot tester at 6 m/s velocity, and 20% weight concentration for 655μm mean particle sizes. The wear specimens were positioned at different orientation angles varying from 15 to 90 deg with respect to the flow direction using the slotted angular plate. The results then compared to existing published results of SS304 at 10% concentration, 4m/s particle velocity and 655 μm as mean particle size. By comparing the results with SS304 material at different material we can observe that Ni-Hard also follows the ductile erosion behaviour.

Table 3- Experimental Procedure

Speed (m/s)	Impact angle (Degree)	Mass Loss (gram)	Erosion Rate (g/g*10-7)
6 m/s	15	0.0184	1.7219
	30	0.0428	4.00548
	45	0.0398	3.72472
	60	0.0375	3.50947
	75	0.0348	3.25679
	90	0.0266	2.48939
8 m/s	15	0.0316	2.2179
	30	0.0468	4.88502
	45	0.0559	3.9234
	60	0.0543	3.811
	75	0.0498	3.49544
	90	0.0376	2.63903

Effect of Impact Angle on Erosion Rate

- The variation of the erosion rate of the Ni-Hard at different orientation angles is presented in the table and figure 3.
- It has been observed from the figure that the erosion rate of the material Ni-hard white cast iron increases with the peaked at 30 deg and then decreases with the further increase in impact angle till 90 deg for both the particle sizes.
- It is evident from the figure 3 by comparing SS304 erosion wear behavior due to maximum wear on 30 deg and then gradually decreasing it can be said that Ni-Hard shows the ductile behavior of erosion.

Effect of Impact Velocity on Erosion Rate

- To investigate the effect of velocity on the slurry erosion of target material, experiments were conducted in the velocity of 6 and 8 m/s.
- First, the measurements were performed at different impact angles and velocities to determine the effect of velocity on the erosion wear dependence of target materials with impact angle.

- The measured erosion rate of the Ni hard at different orientation angles and velocities is presented graphically. It is seen that the increase in velocity increases the erosion rate of the materials.
- The variation in erosion rate of the material shows a similar trend with impact angle at all the velocities. The 30 degree impact angle at which the maximum erosion of the target material occurred is the same for both the velocities.

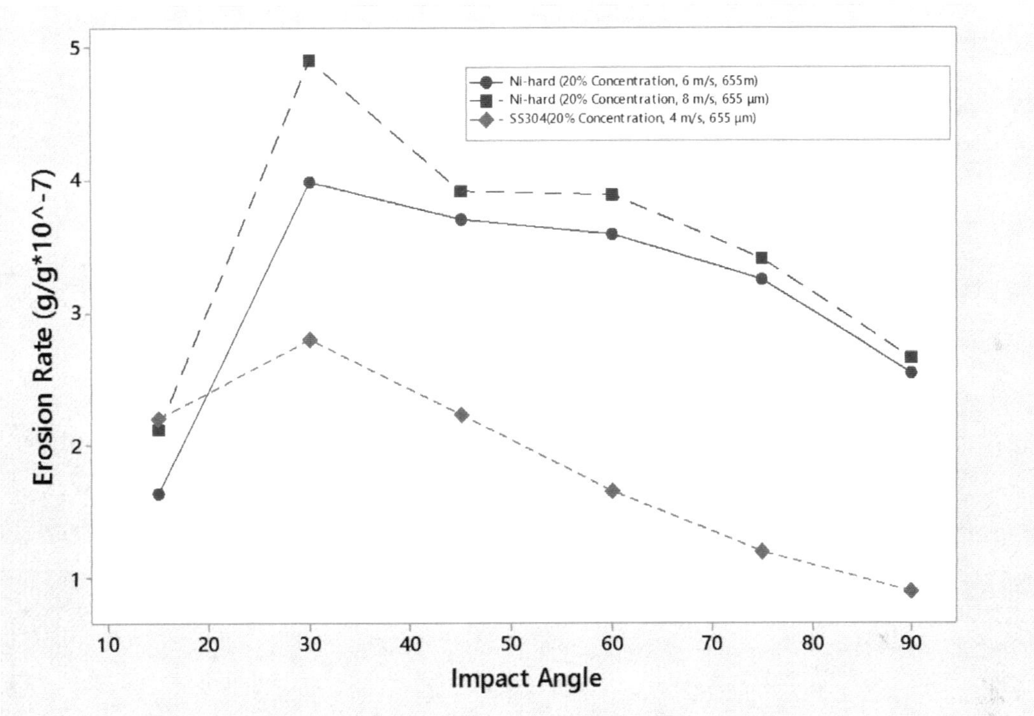

Figure 3: Change of Erosion rate with impact angle

Conclusion

- Erosion wear of Ni hard using slurry erosion pot tester is investigated in with parameters such as particle velocity, impact angle.
- Based on the preliminary experiments it can be said that Ni Hard has ductile type slurry erosion.
- The effect of particle size and slurry concentration is required to study as they are major parameter affecting slurry erosion.
- The experiments shows the relationship of particle velocity and impact angle but further understanding of the parameters and its effects is required to study the dependence on the each parameters
- The erosion of the material increases with impact angle till 30° then the erosion of material decreases gradually.
- The erosion increases the as the velocity of the particle impact increase. Further full scale study required to study velocity variation with respect to other parameters.

References

[1] More Satish R, Nandre Bhushan D,2014,"Development of pot tester to simulate the erosion wear due to solid-liquid mixture",IJRSD,Vol.2

[2] S. Bree, W. F Rosenbrand,., and Gee, A. W. J. de, 1982, "On the erosion resistance in water-sand mixtures of steels for application in slurry pipelines", Hydrotransport 8, BHRA Fluid Engineering, Johannesburg, (S.A.), Paper C3.
[3] Ojala, N., 2012, High Speed Slurry-Pot Type Erosion Wear Tester, The 15th Nordic Symposium on Tribology- NordTrib Trondheim, Norway
[4] Kumar S., Ratol J.S., 2012, "Experimental investigation of erosive wear on the high chrome cast iron impeller of slurry disposal pump using response surface methodology", Materials Engineering - 19, 110-116
[5] Desale G.R., Gandhi B. K., Jain S.C. , 2009, " Particle size effects on the slurry erosion of aluminium alloy,", Wear 266(11), 1066- 1071.
[6] Desale G.R., Gandhi B. K., Jain S.C. , 2005, " Improvement in the design of a pot tester to simulate erosion wear due to solid-liquid mixture", Wear 259(1), 196-202.
[7] Desale G.R., Paul C.P., Gandhi B.K., Jain S.C., 2009, " Erosion wear behavior of laser clad surfaces of low carbon austenitic steel", Wear, Vol. 266, pp. 975-987.
[8] Kumar G, More S R,. Bhatt D V and Menghani J V, Study of Slurry Erosion Wear Behavior of SS-304 in a Slurry Pot test Rig under different operating parameters, 2017, Skit Research Journal, Volume 7; Issue 2: 2017

IOP Conf. Series: Materials Science and Engineering 1004 (2020) 012024 doi:10.1088/1757-899X/1004/1/012024

ABAQUS Implementation to Analyze the Stress Intensity Factor for Laminated Composite Plate

Achchhe Lal [1] and M. B. Vaghela [2*]

[1]Assistant Professor, Department of Mechanical Engineering, S. V. National Institute of Technology, Surat-395007, Gujarat, India.
[2]Ph. D. Research Scholar, Department of Mechanical Engineering, S. V. National Institute of Technology, Surat-395007, Gujarat, India.
*Email: manoj0085@gmail.com

Abstract: The objective of this work is Abaqus implementation to analyze the effect of fibre orientation angle on stress intensity factor for Single Edge Notched (SEN) 3D laminate composite plate under tensile loading by various cases. The Abaqus software is used to analyze the various problems through the extended finite element method (XFEM). The effects of various fibre orientation angles on Sress Intensity Factor (SIF), Von Misses stress and displacement are also investigated. The presented work would be useful for enhancing the use of Abaqus software to analyze the composite structure with crack and other discontinuities under various conditions.

1. Introduction

The Simulate the crack domain into material using conventional finite element methods is quite challenging due to meshing limitations. The extended finite element method (XFEM) is quite useful to solve various discontinuities problems in the material, due to its independent remesh facilities for the crack domain. The crack tip enrichment function is also used in the XFEM to help the better numerical result. The various literatures have been reported on the theories of XFEM which have thrown light on the basic ideas of XFEM.

The material discontinuities like holes, flaw, crack and inclusion are affecting the fracture behaviour of the material. The study about stress, displacement and stress intensity factor will play the major role in the direction of its fracture failure study. It is important in designing the any structure. The investigated crack in a laminated composite plate under tensile loading by implementing ABAQUS by XFEM. In this direction, Sih [1] presented the stress intensity factor evaluation by the energy field theory to measure energy density factor 'S' in the fracture mechanics. The critical value of S gives information on crack initiation and fracture toughness of materials. The strain energy density theory is also capable to handle a variety of mixed-mode crack problems. Ke and Liu [2] presented strain at the crack tip as fracture criteria for ductile material and is compared to J-integral and crack surface opening displacement (CSOD) and it is confirmed that CSOD can be used as one of the fracture criteria but the correlation of these criteria with stress intensity factor has not been established.Chien et al. [3] used the FEM with superposition on the singularity at crack-tip. In this study stress intensity factor at the crack tip of a single edge, the crack plate is calculated and found better accuracy in the results.

In the last decades, researchers have proposed many numerical methods to understand the fracture behaviour of the isotropic plate under different loading conditions. Finite element method (FEM) is broadly used numerical method for many engineering applications, but this method is not so efficient to do desire meshing specially at the crack tip and interior boundary of discontinuities as inclusion, void etc. In FEM, it's necessary to update meshing during the progress of crack and it is time-consuming and costly. In this regard extended finite element method (XFEM) provides extra enrichment functions for the crack face and tip where remeshing is not required during crack growth. The FEM is havening meshing limitation at complex shape, so many researcher had emphasize on evaluate fracture mechanics problem usin the XFEM.

In this direction, Belytschko et al. [4] developed a better methodology to handle discontinuity by XFEM where discontinuity is incorporated without remeshing and this method is successfully implemented to analyze crack growth in the material under loading. The result obtained from this new methodology is very near to the experimental results. Sukumar and Prevost [5] implemented computer program Dynaflow based on finite element method, where the 2-D crack in isotropic and bimaterial plate is analyzed by implementing XFEM in finite element code. This method provided a simple platform where discontinuous fields by the partition of unity framework are incorporated in a standard finite element package.

Bellec and Dolbow [6] presented the modified enrichment in the XFEM to evaluate the SIF at crack tip using ramp function. This study has been carried out for the special case where crack approaches the local nodal spacing. Belytscho et al. [7] presented the importance of XFEM with level set methods and its applications, especially for complicated geometry, crack growth and moving interface. XFEM has increased the capability of the finite element method to solve the fracture mechanics problems which are very difficult for the conventional finite element method. Asadpoure et al. [8] used XFEM for modelling and analysing of 2D orthotropic media where crack–tip displacement fields are added to enrich the finite element with the partition of unity. Here, crack growth is implemented without remeshing and MMSIF is calculated by interaction integral (M-integral).LI and Chen [9] developed formulations for the prediction of the variations of stress intensity factors, using the transformation toughening theory. Mousavi and Sukumar [10] proposed the Gaussian integration method to investigate the integrals weak form by the XFEM. Here, a point elimination algorithm is adopted, where quadrature has a minimal number of Gauss points. Kumar et al. [11] investigated the dynamic response and dynamic SIF at crack-tip in the XFEM. And ramp function is used in the Heaviside function to evaluate the complex problem. Belytschko and Gracie [12] studied a method to calculate Peach-Koehler force which is based on XFEM where discontinuities have been modeled by Volterra dislocation model for a more complex problem. Sukumar et al. [13] presented work on modelling holes and inclusions by level set in XFEM, where they described the modelling the internal boundaries where local enrichment functions in the generalized finite element method for modelling corners in two dimensions are used. Wiroj et al. [14] presented an adaptive finite element method to determine stress intensity factors K_1 and K_2 of the polycarbonate edge crack plate with with considering void and inclusion using FEM and found better results as compared to photoelastic method. Jiang et al. [15] investigated the crack growth with the interaction of various discontinuities like crack, void, hard and soft inclusion by XFEM. Natrajan et al [16] presented a numerical analysis of inclusion crack interaction by XFEM where both inclusion and crack are modelled within the XFEM framework and found better accuracy and flexibility in the results. Sharma [17] simulated the material discontinuities interaction effect on fracture behaviour by XFEM.. Stress intensity factor (SIF) is calculated by the interaction integral method and effect of crack orientation under mechanical loading considering hard inclusion.

Ismael Rivero Arévalo et al [18] is evaluated the Abaqus XFEM capabilities to analyze the crack growth of aeronautical structure and it could be a beneficiary in industrial application. M. Al-Khalil et al [19] used the Abaqus for investigating the damage reduction in composite structure during bird impact. Kaman [20] investigated the effect of fibre orientation angle on the laminated composite plate through the experimental study. He has also used Ansys for numerical study. Lal et al [21] presented the study of stochastic analysis of the SIF for SEN laminated composite plate using Matlab. Xu et al [22] implemented the Abaqus software for crack problem through the extended finite element method.

From the above literature study, it is observed that researchers are focused on Abaqus implementation of composite structure, fracture analysis through the XFEM with the interaction of crack under different loading condition.

In this present work main focus is Abaqus implementation to analyze the stress intensity factor, von misses stress and displacement for SEN laminated composite plate under tensile loading. It is investigated the SIF K_I, von misses stress and displacement for various case studies, like various fibre orientation angle under tensile loading.

2. Mathematical Formulation

2.1. Calculation of critical stress intensity factor based on the experimental results

The stress intensity factor for the Mode I can be written as,

$$(K_I) = \sigma f\left(\frac{b}{w}\right)\sqrt{\pi b} \tag{1}$$

where, σ is the applied normal stress and $f\left(\frac{b}{w}\right)$ is a geometrical factor. K_I Provides the severity of the crack tip environment and it is logical to characterize resistance to fracture by a critical value that is K_{IC} When the applied normal stress reaches the failure stress, the stress intensity factor K_I becomes $(K_{IC})_{EXP}$, the critical stress intensity factor, which is taken as the fracture toughness of the composite material. The value of critical stress intensity factor (K_{IC}) is given by,

$$(K_{IC}) = \sigma_f f\left(\frac{b}{w}\right)\sqrt{\pi b} \tag{2}$$

where, σ_f is the applied maximum normal stress or failure stress. The geometrical factor for single edge notched specimens [20] is,

$$f\left(\frac{b}{w}\right) = 1.12 - 0.231\left(\frac{b}{w}\right) + 10.55\left(\frac{b}{w}\right)^2 - 21.72\left(\frac{b}{w}\right)^3 + 30.39\left(\frac{b}{w}\right)^4 \tag{3}$$

2.2. Numerical study of Displacement Correlation Method (DCM) [20]

The DCM is used to evaluate the fracture toughness from the failure load. For singular finite element model at the crack tip Figure 1 the crack opening displacement (COD) and the crack sliding displacement (CSD) are given by,

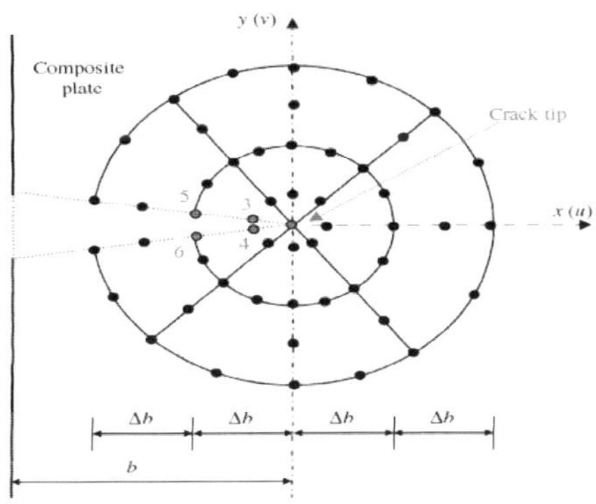

Figure 1: Finite element model at the crack tip

$$COD=\sqrt{\frac{r}{\Delta b}}\left(4\Delta V_{34}-\Delta V_{56}\right)=\sqrt{\frac{r}{\Delta b}}\left[4\left(V_3-V_4\right)-\left(V_5-V_6\right)\right]$$

$$CSD=\sqrt{\frac{r}{\Delta b}}\left(4\Delta u_{34}-\Delta u_{56}\right)=\sqrt{\frac{r}{\Delta b}}\left[4\left(u_3-u_4\right)-\left(u_5-u_6\right)\right] \tag{4}$$

where, b is the crack length and Δb is the characteristic length of the quarter point crack tip elements, Δu_{34}, Δu_{56} and Δv_{34}, Δv_{56} are the relative displacements with respect to crack-tip in the x, y-directions, respectively. Displacements at the crack tip can be written as,

$$u=K_I\sqrt{\frac{2r}{\pi}}\,\mathrm{Re}\left[\frac{i}{\mu_1-\mu_2}\left(\mu_1 q_2-\mu_2 q_1\right)\right]+K_{II}\sqrt{\frac{2r}{\pi}}\,\mathrm{Re}\left[\frac{i}{\mu_1-\mu_2}\left(q_2-q_1\right)\right]$$

$$v=K_I\sqrt{\frac{2r}{\pi}}\,\mathrm{Re}\left[\frac{i}{\mu_1-\mu_2}\left(\mu_1 p_2-\mu_2 p_1\right)\right]+K_{II}\sqrt{\frac{2r}{\pi}}\,\mathrm{Re}\left[\frac{i}{\mu_1-\mu_2}\left(p_2-p_1\right)\right] \tag{5}$$

where, r is the distance from the crack tip along the x axis, μ_1 and μ_2 are the eigenvalue of the field equation with positive imaginary part of te orthotropic body calculated from,

$$S_{11}\mu^4+(2S_{12}+S_{66})\mu^2+S_{22}=0 \tag{6}$$

where, S_{ij} (i, j =1, 2, 6) are the compliance coefficients of the composite material

$$\varepsilon_i=S_{ij}\sigma_j \text{ (i, j=1, 2, 6)} \tag{7}$$

Here ε_{ij} (i, j=1, 2, 6) is the strain and σ_j is the stress vector p_k and q_k (k = 1, 2) are given by

$$p_K=S_{11}\mu_K{}^2-S_{16}\mu+S_{12}$$

$$q_K=S_{12}\mu_K{}^2+\frac{S_{22}}{\mu_k}-S_{26} \tag{8}$$

By using Eqs. (5-8), formulas for pure Mode I stress intensity factor (K_I) and Mode II stress intensity factor (K_{II}) can be obtained:

$$K_I=\frac{1}{4}\sqrt{\frac{2\pi}{\Delta b}}\frac{\left(D\left(4\Delta u_{34}-\Delta u_{56}\right)-B\left(4\Delta v_{34}-\Delta v_{56}\right)\right)}{AD-BC} \tag{9}$$

$$K_{II}=\frac{1}{4}\sqrt{\frac{2\pi}{\Delta b}}\frac{\left(A\left(4\Delta v_{34}-\Delta v_{56}\right)-C\left(4\Delta u_{34}-\Delta u_{56}\right)\right)}{AD-BC} \tag{10}$$

where A, B, C & D are given by,

$$A=\mathrm{Re}\left[\frac{i}{\mu_1-\mu_2}\left(\mu_1 p_2-\mu_2 p_1\right)\right] \tag{11}$$

$$B=\mathrm{Re}\left[\frac{i}{\mu_1-\mu_2}\left(p_2-p_1\right)\right] \tag{12}$$

$$C=\mathrm{Re}\left[\frac{i}{\mu_1-\mu_2}\left(\mu_1 q_2-\mu_2 q_1\right)\right] \tag{13}$$

$$D=\mathrm{Re}\left[\frac{i}{\mu_1-\mu_2}\left(q_2-q_1\right)\right] \tag{14}$$

2.3. Analysis of Stress Intensity Factor Laminate Composite Plate by Abaqus through the XFEM

To perform the analysis of stress intensity factor, the extended finite element analysis is used. The Abaqus software is used to perform the extended finite element analysis (XFEA).

Input options in the Abaqus for material properties are E (Young's modulus) and v (Poisson's ratio) for homogeneous linear isotropic material, E_x and E_y (Young's moduli in the x- and y-directions, respectively), v_{xy} (Poisson's ratio) and G_{xy} (shear modulus) for homogeneous linear orthotropic material on the plane. This element has higher order interpolation functions which valid to more accurate results. Material properties of laminates are calculated by using the classical laminated plate theory and are supplied to the program [20]. In a symmetric laminated composite plate stress–strain relation is.

$$\begin{Bmatrix} N_X \\ N_Y \\ N_{XY} \end{Bmatrix} = \begin{bmatrix} A_{11} & A_{12} & A_{16} \\ A_{21} & A_{22} & A_{26} \\ A_{16} & A_{26} & A_{66} \end{bmatrix} \begin{Bmatrix} \varepsilon_x \\ \varepsilon_y \\ \varepsilon_{xy} \end{Bmatrix} \tag{15}$$

where,

$$N_X = \sigma_X t,\ N_Y = \sigma_y\ ,\ N_Z = \sigma_y \quad \text{and} \quad A_{ij} = \sum_{k=1}^{N} \overline{Q}_{ij}^{(k)} h_k \tag{16}$$

In which $\overline{Q}_{ij}^{K}$ are reduced stiffnesses and h_k the thickness of the k_{th} layer. The longitudinal (E_x), transverse (E_y), shear (G_{xy}) moduli, and Poisson's ratio (υ_{xy}) of laminated composite plates are respectively equal to

$$E_x = \frac{N_x / t}{\varepsilon_x}, E_y = \frac{N_y / t}{\varepsilon_y}, G_{xy} = \frac{T_{xy} / t}{\gamma_{xy}}, \upsilon_{xy} = \frac{-\varepsilon_y}{\varepsilon_x} \tag{17}$$

And it can be derived by inverting the 3 x 3 stiffness matrix [A].

$$[A]^{-1} = \begin{bmatrix} A_{11}^* & A_{12}^* & A_{16}^* \\ A_{12}^* & A_{22}^* & A_{26}^* \\ A_{16}^* & A_{26}^* & A_{66}^* \end{bmatrix} \tag{18}$$

Longitudinal modulus (E_x) for the laminated composite plates as,

$$E_x = \frac{Nx/t}{\varepsilon_x} = \frac{1}{A_{11}^* t} = \left(\frac{A_{16}^2 A_{22} - 2A_{12}A_{16}A_{26} + A_{11}A_{26}^2 + A_{12}^2 A_{66} - A_{11}A_{22}A_{66}}{A_{26}^2 - A_{22}A_{66}} \right) / t \tag{19}$$

E_y and G_{xy} expressed as,

$$E_y = \frac{Ny/t}{\varepsilon_y} = \frac{1}{A_{22}^* t} = \left(\frac{A_{16}^2 A_{22} - 2A_{12}A_{16}A_{26} + A_{11}A_{26}^2 + A_{12}^2 A_{66} - A_{11}A_{22}A_{66}}{A_{16}^2 - A_{11}A_{66}} \right) / t \tag{20}$$

$$G_{xy} = \frac{Txy/t}{\gamma_{xy}} = \frac{1}{A_{66}^* t} = \left(\frac{A_{16}^2 A_{22} - 2A_{12}A_{16}A_{26} + A_{11}A_{26}^2}{A_{12}^2 - A_{11}A_{22}} + A_{66} \right) / t \tag{21}$$

E_{xy} can be written as,

$$E_{xy} = \sqrt{E_X * E_y} \tag{22}$$

By using Eq. (19-21) respectively, the Poisson's ratio (υ_{xy}) can be calculated from $\frac{-A_{12}}{A_{11}}$. Finally, the compliance coefficients of composite material S_{ij} (i, j = 1, 2, 6) for plane element can be written as

$$S_{ij} = \begin{bmatrix} S_{11} & S_{12} & S_{16} \\ S_{12} & S_{22} & S_{26} \\ S_{16} & S_{26} & S_{66} \end{bmatrix} = \overline{S}_{ij} = \begin{bmatrix} \frac{1}{E_x} & \frac{-\nu_{xy}}{E_y} & 0 \\ \frac{-\nu_{xy}}{E_y} & \frac{1}{E_y} & 0 \\ 0 & 0 & \frac{1}{G_{xy}} \end{bmatrix} \tag{23}$$

Eq. (6) can be written by using calculated material properties (E_x, E_y, G_{xy} and m_{xy}) of laminated plates,

$$\overline{S_{11}}\mu^4 + \left(2\overline{S_{12}} + \overline{S_{66}}\right)\mu^2 + \overline{S_{22}} = 0$$

Material properties required as an input are taken from following Table 1. For every lamina configuration the properties are different in X and Y axis. The properties required for orthotropic material are E_x, E_y, E_z= Young's modulus of elasticity in x, y, z direction respectively (MPa), G_{xy}, G_{yz}, G_{xz}= Shear modulus in xy, yz, xz plane respectively (MPa) and Poison's ratios in xy, yz, xz plane.

Table 1: Laminate Material Properties

	Kaman [20]				Present
Fibre Orientation $[0°/\theta°]_s$	E_x (Mpa)	E_y (Mpa)	G_{xy} (Mpa)	ν_{xy}(-)	E_{xy} (Mpa)
[0°/15°/15°/0°]	1.191 x 10^5	1.076 x 10^4	6.514 x 10^3	0.511	3.580 x 10^4
[0°/30°/30°/0°]	9.771 x 10^4	1.143 x 10^4	1.184 x 10^4	0.546	3.342 x 10^4
[0°/45°/45°/0°]	9.238 x 10^4	1.349 x 10^4	1.050 x 10^4	0.474	3.530 x 10^4
[0°/60°/60°/0°]	9.129 x 10^4	1.922 x 10^4	6.572 x 10^3	0.329	4.189 x 10^4
[0°/75°/75°/0°]	9.180 x 10^4	4.007 x 10^4	4.311 x 10^3	0.139	6.065 x 10^4
[0°/90°/90°/0°]	9.230 x 10^4	9.230 x 10^4	3.638 x 10^3	0.038	9.230 x 10^4

3. Result and Discussion

3.1. Validation study for Single Edge Notched (SEN) laminated composite plate using Abaqus

The SEN 3D laminated composite plate under tensile loading is considered for a validation study. To analyze the SIF KI for SEN composite plate, define the problem model as shown in Figure 2.

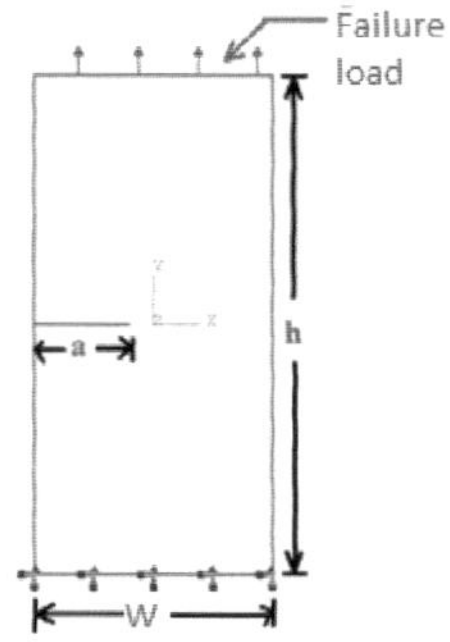

Figure 2: The SEN laminated composite plate under tensile loading

To analyze the SIF K_I for 3D laminated composite plate used the material properties as per Table 1 and for geometry dimension and loading as per Table 2. Average failure load is applied on the top surface of plate and applied boundary condition at the bottom surface as fixed as shown Figure 2. Abaqus 6.13 is used to analyze the SIF K_I, Von misses stress and displacement for SEN laminated composite plate. For meshing chosen *Hex* element shape, *sweep* technique and *medial* axis algorithm. Total numbers of nodes are 5238 and total number of elements is 1540.

The results are presented in three main sections: (a) experimental results $(K_{IC})_{EXP}$ with using the empirical relation for SEN specimens, (b) numerical simulation results $(K_{IC})_{PLANE}$ and $(K_{IC})_{SHELL}$ of two applied elements *Plane82* and *Shell99* from experimental failure load (c) analysis of $(K_{IC})_{ABAQUS}$ from experimental failure load, respectively.

Table 2: Experimental failure loads for different fiber orientation angle

Kaman [20]					Present
Fibre Orientation Angle θ (°)	Crack length *b* (mm)	Width of plate *w* (mm)	Total laminate thickness *t* (mm)	Average failure load (N)	Failure stress (MPa)
15	9	18	1.235	6516.112	293.077
30	9	18	1.235	5145.807	231.480
45	9	18	1.235	4722.581	212.442
60	9	18	1.235	5229.024	235.224
75	9	18	1.235	4193.829	188.656
90	9	18	1.235	3680.767	165.577

As per experiment result the maximum average failure load is 6516.112 N for $[0°/15°]_S$ lamina configuration. Generally, failure load tends to decrease with the increasing fiber orientation angle θ. When θ reaches 90°, the minimum failure load is obtained (3680.767 N). The SIF is getting the maximum for $[0°/15°]_S$ lamina configuration in experimental, FEA through Ansys and by Abaqus and minimum for $[0°/90°]_S$ lamina configuration as shown in Figure 3. As shown in Figure 3, Abaqus results are getting quite nearer to the experimental result.

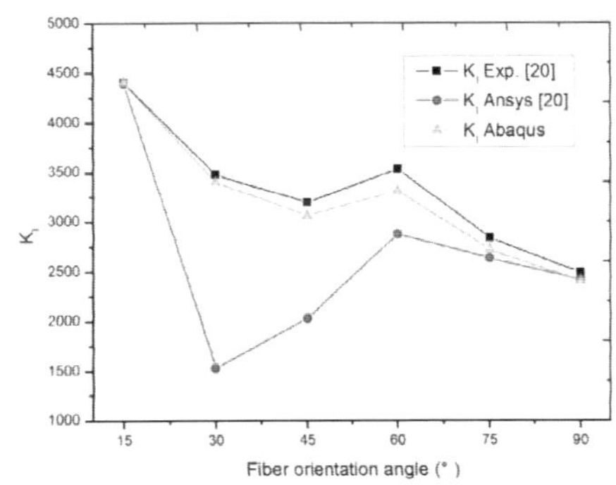

Figure 3: Validation of SIF K_I of SEN laminated composite plate for various fibre orientation angle (θ)

Figure 4 and Figure 5 shows the Von Mises stress and displacement of the SEN laminated composite plate respectively from Abaqus 6.13. The maximum value of Von Mises stress and displacement are getting at fibre orientation angle 15° and decrease as per for fibre orientation angle 15°, 30°, 45°, 60°, 75°, 90° increase respectively.

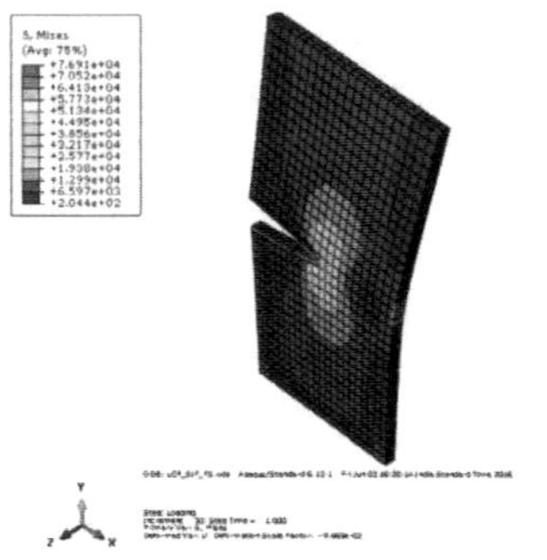

Figure 4: Von misses stress of SEN laminated composite plate for fibre orientation angle (15°)

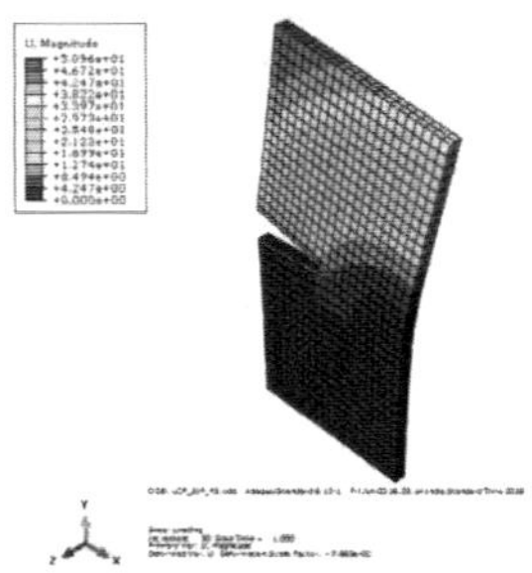

Figure 5: Displacement of SEN laminated composite plate for fibre orientation angle (15°)

3.2. Analyze the Single Edge Notched Composite Plate for variable fibre orientation angle using Abaqus

To analyze the SIF K_I for SEN 3D laminated composite plate defined the problem model as shown in Figure 2. The material properties are considered as per Table 1 and for the geometry dimension and loading as per Table 2. Average failure load is applied on the top surface of plate and applied boundary condition at the bottom surface is fixed as shown Figure 2. Abaqus 6.13 is used to analyze the SIF K_I for SEN 3D laminated composite plate, in meshing chosen *Hex* element shape, *sweep* technique and *medial* axis algorithm. Total numbers of nodes are 2944 and total number of elements is 630.

3.2.1. The SIF K_I result analysis for variable fibre orientation angle.

The results are presented in four main sections: (a) experimental results $(K_{IC})_{\mathrm{EXP}}$ with using the empirical relation for SEN specimens, (b) numerical simulation results $(K_{IC})_{\mathrm{PLANE}}$ and $(K_{IC})_{\mathrm{SHELL}}$ of two applied elements *Plane82* and *Shell99* from experimental failure load (c) analysis of $(K_{IC})_{\mathrm{ABAQUS}}$ from experimental failure load and (d) analysis of $(K_{IC})_{\mathrm{MATLAB}}$ from experimental failure load, respectively.

As per author considered result of the maximum average failure load is 6516.112 N for $[0°/15°]_S$ lamina configuration. Generally, failure load tends to decrease with the increasing fiber orientation angle θ. When θ reaches 90°, minimum failure load is obtained (3680.767 N). The SIF K_I is getting maximum for $[0°/15°]_S$ lamina configuration in experimental, FEA through Ansys, MATLAB and Abaqus by XFEM and minimum for $[0°/90°]_S$ lamina configuration as shown in Figure 6. The Abaqus results are getting very nearer to Matlab result and more improved result as compare to Ansys result.

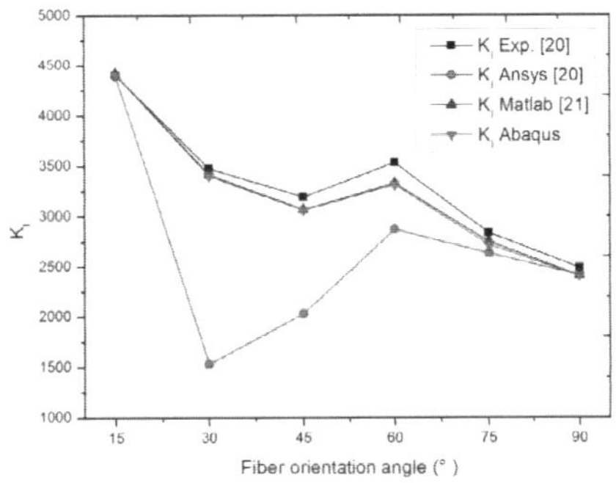

Figure 6: SIF K_I of SEN Laminated Composite Plate for various fibre orientation angle (θ)

3.2.2. The von misses stress and displacement result analysis for variable fibre orientation angle .
The The result of Von misses stress and displacement for the SEN laminated composite plate are getting as shown Figure 7 and Figure 8 respectively from Abaqus 6.13. The maximum value of Von misses stress and displacement are getting at fibre orientation angle 15° and decrease as per for fibre orientation angle 15°, 30°, 45°, 60°, 75°, 90° increase respectively.

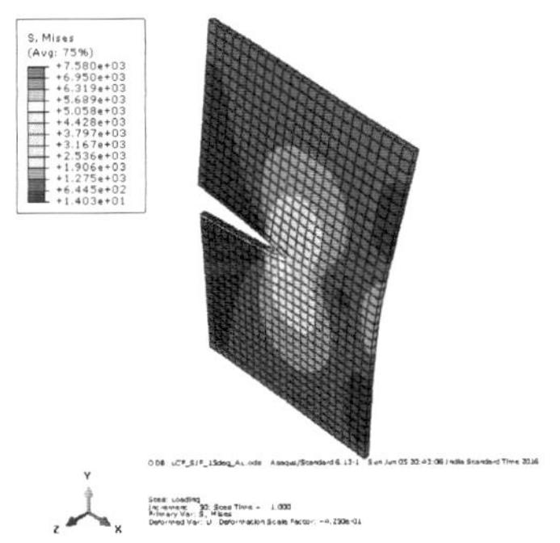

Figure 7: Von misses stress of SEN laminated composite plate for fibre orientation angle (15°)

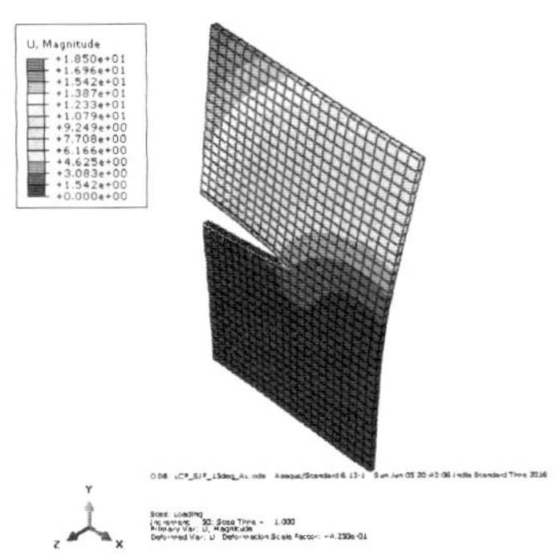

Figure 8. Displacement of SEN Laminated Composite Plate for fibre orientation angle (15°)

4. Conclusions

The present work is analyzed, the Abaqus implementation to analyze the effect of fibre orientation angle on stress intensity factor for the SEN 3D laminate composite plate under tensile loading. The following evaluations are found during this study;

- The SIF K_I results through the Abaqus software are getting nearer to experimental results [20] for taken case.
- The SIF K_I results through the Abaqus software are getting very nearer to the Matlab results and improved as compared to Ansys software result for taken case.
- The SIF K_I is getting maximum for $[0°/15°]_S$ lamina configuration and minimum for $[0°/90°]_S$. And found decreasing nature by increase the fibre orientation angle increased.
- The maximum value of Von misses stress and displacement are getting at fibre orientation angle 15° and decrease as per for fibre orientation angle 15°, 30°, 45°, 60°, 75°, 90° increase respectively.
- Through this study, it is found that the Abaqus implementation through the XFEM results are improved and would be more applicable for industrial application as well in research

References

[1] Sih G C 1973 *Eng. Fract. Mech.* **5,** 1037-1040.
[2] Ke J S and Liu H W 1973 *Eng. Fract. Mech.* **5,** 187-202.
[3] Chien W Z, Xie Z C, Gu Q L, Yang Z F and Zhou C T 1982 *Eng. Fract. Mech.* **16,** 95-103.
[4] Belytschko T, Chen H, J Xu, and Zi G 2003 *Int. J. Numer. Meth. Engng* **58,** 1873-1905.
[5] Sukumar N and Prevost J H 2003*Int J Solids Struct.* **40,** 7513-753.
[6] Bellec J and Dolbow J E 2003 *Commun. Numer. Meth. Engng.* **19,** 921–932.
[7] Belytschko T, Gracie R and Ventura G A 2005 *Int. J. of Fract.* **131,** 124–129.
[8] Asadpoure A, Mohammadi S and Vafai A 2006 *Finite Elements in Analysis and Design.* **42,** 1165-75
[9] LI Z and Chen Q 2002 *Int. J. Fract.***118,** 29–40.
[10] Mousavi S E and Sukumar N 2010 *Comput. Methods Appl. Mech. Engrg.* **199,** 3237-3249.
[11] Kumar S, Singh I V, Mishra B K and Singh A 2016 *International Journal of Impact Engineering* **87,** 198-211.
[12] Belytschko T and Gracie R 2007 *International Journal of Plasticity* **23,** 1721-1738.
[13] Sukumar N, Chop D L, Moes N and Belytscko T 2001 *Apll. Mech.* **190,** 6183-6200.
[14] Limtrakarn W and Dechaumphai P 2011 *T Can Soc. Mech. Eng.* **35,** 355-368.
[15] Jiang S, Du C and Gu C 2014 *Struct. Eng. Mech.* **49-5,** 597-618.
[16] Natarajan S, Kerfriden P, Mahapatra D R and Bordas S P A 2014 *International Journal for Computational Methods in Engineering Science and Mechanics* **15,** 26-32.
[17] Sharma K 2014 *J. Solid Mech.* **6-4,** 410-421.
[18] Arevalo I R and Martin J G E 2015 *SIMULIA Community Conference Proceedings* 23-36.
[19] Khalil M A, Rigby R and Kirtil E 2015 *SIMULIA Community Conference Proceedings* 48-62.
[20] Kaman M O 2011 *Eng. Fract. Mech.* **78,** 2531-2534.
[21] Lal A and Kapania R K 2013 *54th AIAA/ASME/ASCE/AHS/ASC Structures, Structural Dynamics and Materials Conference* **4,** 2653-2675.
[22] Xu Y and Yuan H 2009 *Eng. Fract. Mech.* **76,** 165–181.

Buckling Response Analysis of Laminated Plates Subjected to Localised Bi-axial In-plane Compressive Loading

Achchhe Lal [1], Anant Parghi [2], Anil Kumar Mahto [3], Rahul Kumar [4]

[1] Assistant Professor, Department of Mechanical Engineering SVNIT, Surat, India.
[2] Associate Professor, Department of Mechanical Engineering SVNIT, Surat, India.
[3] Assistant Professor, Department of Mechanical Engineering SVNIT, Surat, India.
[4] Research Scholars, Department of Mechanical Engineering SVNIT, Surat., India.
*E-Mail: lalachchhe@yahoo.co.in

Abstract: Laminated composite plates are well known structural member since last four decades because of its high stiffness and strength to weight ratio. Usually this type of structural members is subjected to partial biaxial loading during their functioning. For better safety and stability of structure, analysis of these structural member in that condition of loading is necessary. In this present analysis buckling response of laminated composite plates under biaxial partial in-plane loading is examined and effect of various parameters like aspect ratio, stacking sequence, fiber orientation and plate thickness ratio of nondimensionalised critical buckling load (NCBL) are revealed. A MATLAB program is developed higher order shear deformation theory based on finite element method for buckling analysis of laminated composite plate. Efficiency of the program is checked by comparing the result of the present work with that of already published which shows good favour of later one.
Keywords: Laminated Composite Plate, biaxial loading

1. Introduction

Application of laminated composite plates as a structural member is very common in weight sensitive area like aerospace vehicles, submarines, bridges etc. due to its outstanding stiffness and strength to weight ratio. These plates are usually subjected partial in-plane loading by their adjacent structural member. Therefor for better safety and reliability proper analysis of these structural member are required.

Many researches are available for buckling behaviour analysis of plates under in-plane edge loadings. Osama et al. [1-2] done buckling analysis of laminated composite plate under biaxial in-plane loading. [1] used classical laminated plate theory (CLPT) and find the effect of stacking sequence, aspect ratio and modulus ratio on buckling load. [2] done survey of different plate theories and method to examine buckling behaviour of laminated plate, beside this the author achieved various other objectives like developed theoretical model to predict buckling load, studied about finite element based technique for behaviour analysis of laminated plates under buckling, accuracy of the developed technique, influence of coupling parameter of bending and twisting on buckling response and generated results based on CLPT. Libove [3] studied buckling of orthotropic under biaxial loading when edges are simply supported and found that plates buckle with half waves in the two-principal direction. Shukla et al. [4] used first order shear deformation theory with von-Karman-type nonlinearity to estimated buckling load of laminated composite plates subjected to uniaxial and biaxial loadings and revealed the effect of support conditions, aspect ratio, stacking sequence, number of layers and properties of material on buckling loads. Tung and Surdenas [5] analysed orthotropic plate for buckling response when plate subjected to biaxial loading. Wnag and Wang [6] obtained buckling load for plate under uniaxial or biaxial compressive concentrated load using differential quadrature method and compared the results with that of existing solution.

Sometimes structural members are not subjected to load to whole of the edges besides that subjected to partial edge loading by their adjacent structural member. Analysis of these structural members in that condition of locally applied load introduces more complexity.
Spencer and Surjanhata [7] simplified buckling criteria for partial uniaxial loading and compared the results with numerical solutions. Srivastava et al. [8] studied stability of stiffened plates under localised edge loading using finite element technique and find the effect of aspect ratio, boundary conditions and different non-uniform loadings on NCBL and frequencies. Baker and Pavlovic [9] used Mahtieu's exact solution for evaluation of buckling coefficients for subjected to locally distributed load and concentrated load.

From the literature survey it is observed that stability of laminated composite plate under biaxial partial in-plane compressive load needs to be analysed for efficiently used as structural member. In the present work buckling response analysis of laminates under biaxial localised in-plane edge loading is done. Two cases for different position of load application is considered here and their effect on NCBL is investigated. The effect of increment of extent of loading on same is also studied.

2. Problem formulation

Buckling response analysis of laminated composite plate having dimension 1 x 1 x 0.1 subjected to biaxial partial in-plane loading is done with finite element based MATLAB program using nine noded isoperimetric quadrilateral element.

Table 1: Mechanical properties of material

E_1 (GPa)	E_2 (GPa)	$\nu_{12} = \nu_{21}$	ν_{23}	G_{12} (GPa)
82.64	9.8034	0.3	0.0356	3.8545

2.1. *Displacement field*

In the present study the displacement field can be given as,

$$\begin{aligned} u_1 &= u_0 + z\psi_x + z^2\xi_x, \\ u_2 &= v_0 + z\psi_y + z^2\xi_y, \\ u_3 &= w_0, \end{aligned} \tag{1}$$

2.2 *Strain-displacement expression*

$$\begin{aligned} &\varepsilon_x = \frac{\partial u_1}{\partial x}, \quad \varepsilon_y = \frac{\partial u_2}{\partial y}, \quad \varepsilon_{xy} = \frac{\partial u_1}{\partial y} + \frac{\partial u_2}{\partial x} \\ &\varepsilon_{yz} = \frac{\partial u_3}{\partial y} + \frac{\partial u_1}{\partial z}, \quad \varepsilon_{zx} = \frac{\partial u_1}{\partial z} + \frac{\partial u_3}{\partial x} \end{aligned} \tag{2}$$

2.3. *Stress-strain relations*

Constitutive relation for the present work can be given as,

$$\begin{bmatrix} \sigma_{xx} \\ \sigma_{yy} \\ \tau_{xy} \\ \tau_{yz} \\ \tau_{zx} \end{bmatrix} = \left[\overline{Q_b}\right] \begin{bmatrix} \varepsilon_{xx} \\ \varepsilon_{yy} \\ \gamma_{xy} \\ \gamma_{yz} \\ \gamma_{zx} \end{bmatrix} \tag{3}$$

Where $\left[\overline{Q_b}\right]$ is reduced constitutive matrix.

Boundary condition applied in the present study is given as,

$$v = 0,\ w = 0,\ \psi_y = 0,\ \xi_y = 0; \qquad at \quad x = 0, a$$

$$u = 0,\ w = 0,\ \psi_x = 0,\ \xi_x = 0; \qquad at \quad y = 0, b$$

2.4. *Finite element approach*

Displacement vector can be given as,

$$U^e = \begin{bmatrix} u_1 & v_1 & u_2 & v_2 \dots & u_n & v_n \end{bmatrix}^T \tag{4}$$

$$U_1 = \sum_{i=1}^{N} N_i^e u_i \quad and \quad U_1 = \sum_{i=1}^{N} N_i^e v_i \tag{5}$$

Where, N_i^e is elemental shape function.

Strain vector can be given as

$$\{\varepsilon\} = [L]\{U^e\} \tag{6}$$

The stiffness matrix of laminated plate can be given as,

$$[K_1] = \int_{-1}^{+1}\int_{-1}^{+1} [L_c]^T [\bar{Q}][L_c]|J|\, d\xi\, d\eta, \tag{7}$$

Where, $[L_c]$ is kinematic matrix, $|J|$ is determinant of Jacobian matrix and $[\bar{Q}]$ is transformed constitutive matrix.

The geometric stiffness matrix of plate may be written as,

$$[K_{gc}] = \int_{-1}^{+1}\int_{-1}^{+1} [L_{gc}]^T [N_t][L_{gc}]|J|\, d\xi\, d\eta, \tag{8}$$

where, $[L_{gc}]$ is geometric kinematic matrix and $[N_0]$ is load vector.

$$\begin{aligned} &[N_t] = [N_{xx} \quad N_{xy};\ N_{xy} \quad N_{yy}], \\ &\text{Where, } [N_{xx}] = 1/\text{thickness}, \\ &[N_{xy}] = 0 \text{ and } [N_{yy}] = 1/\text{thickness} \end{aligned} \tag{9}$$

For partial edge loading $[N_x] = C_x$, where, C_x is x-coordinates of nodes of edge on which load is applied.

$$[L_{gc}] = \begin{bmatrix} 0 & 0 & \partial N_i/\partial x & 0 & 0 & 0 & 0 \\ 0 & 0 & \partial N_i/\partial x & 0 & 0 & 0 & 0 \end{bmatrix}, \tag{10}$$

2.5. *Governing equation for buckling*

The governing equation for buckling response of plate acted up on by in-plane partial loading may be given as,

$$\left[[K_l] - \lambda_{cr}[K_g]\right]\{\vee\} = 0, \tag{11}$$

NCBL for partial in-plane loading may be given as,

$$\lambda_{cr} = \frac{\lambda c b^2}{E_2 h^3}, \tag{12}$$

Where 'c' is percentage of width of plate on which partial load is acted upon.

3. Result and Discussion

NCBL of laminated composite plate subjected to biaxial partial in-plane loading for different position of loading along the edge are evaluated. Table 1 Shows validation study for NCBL for a simply supported laminated plate (0/90/0/90/0) under a/h=20 and a/h=100. It is revealed from the Table 2, that outcomes are in good agreement with the published result.

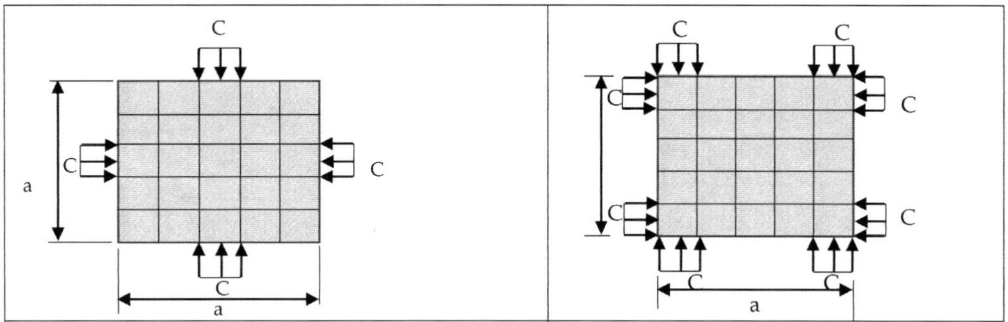

(a) Case-I (b) Case-II

Figure 1: Biaxial partial in-plane compressive loading on 20 % of edge for two different position (a) Case-I and (b) Case-II.

Table 2: Validation of non-dimensional critical buckling load for a simply supported laminated composite plate with stacking sequence (0/90/0/90/0) subjected to in-plane partial edge loadings.

c/a	a/h = 20		a/h =100	
	Present	Ref. [7]	Present	Ref. [7]
0.2	15.2413	16.287	18.4578	19.563
0.4	17.3145	18.417	21.6318	21.590
0.6	20.6718	21.810	25.2101	25.110
0.8	24.8345	26.487	31.7152	30.205
1.0	30.6587	31.958	32.5478	35.962

Table 3: Variation of NCBL with aspect ratios and extent of edge loadings of simply supported laminated composite (0\90\0\90\0) with a/h=100 subjected to partial bi-axial compressive in-plane edge loading at different position.

c/a	NCBL				
	a/b=1	**a/b=1.5**	**a/b=2**	**a/b=2.5**	**a/b=3**
Case-I					
0.2	10.974	7.2099	5.869	4.3097	3.414
0.4	8.3357	5.5700	4.604	3.3970	2.705
0.6	10.974	7.2099	5.869	4.3097	3.414
0.8	9.6664	6.4290	5.291	3.8988	3.100
1.0	13.629	8.7099	6.980	5.0807	4.009
Case-II					
0.2	5.751	3.8004	3.110	2.2872	1.815
0.4	11.21	7.0919	5.676	4.1249	3.257
0.6	8.723	5.6510	4.582	3.3511	2.654
0.8	10.26	6.6931	5.412	3.9654	3.134
1.0	13.62	8.7099	6.980	5.0807	4.009

Table 3 shows that buckling load is decreases with increase in the length of the plate. Plate with aspect ratio 3 shows 68% less buckling load than that in case of aspect ratio is one. Plate with loading at whole edge i.e. uniform loading requires more load to buckle than the plate when subjected to nonsymmetric loading. When loading is applied at centre as in case-I plate shows 47.6% more buckling load than that in case-II loading condition.

Table 4: Effect of span to thickness ratio on NCBL of simply supported square laminated composite plate (0\90\0\90\0) under bi-axial in-plane partial (20% of edge) edge loading at different positions.

	NCBL

a/h	Case-I	Case-II
10	6.3823	3.3768
20	8.8273	4.6383
30	9.5447	5.0077
40	9.9070	5.1950
50	10.1607	5.3266

Table 4 Reveals the effect of plate span to thickness ratio of NCBL of laminated composite plate (0/90/0/90/0) under biaxial in-plane loading when loading extent is 20% of the edge of plate. With increasing thickness ratio, NCBL increases and shows 37.18% increment in buckling load when thickness ratio changes from 10 to 50 in loading case-I while only 63.6% as in case-II.

Table 5 shows variation of NCBL with fiber orientation, as fiber orientation increases from 00 to 450 NCBL increases after that from 450 to 900 it starts decreasing because fibers have highest strength along longitudinal direction.

Table 5: Effect of fiber orientation along with the position of loading on NCBL of square laminated composite plate with a/h=100 under biaxial partial compressive in-plane edge loading at 20 % of width of plate.

Fiber Orientation	NCBL	
	Case-I	Case-II
90/-90/90/-90/90	4.4629	2.3598
60/-60/60/-60/60	13.1983	6.9636
45/-45/45/-45/45	17.1311	8.9966
30/-30/30/-30/30	15.7264	8.2445
0/0/0/0/0	10.8721	5.6960

Table 6. Variation of NCBL with stacking sequence and biaxial partial (20 % of edge) in-plane loading position for a simply supported square laminated composite plate with a/h=100.

Stacking sequence	NCBL	
	Case-I	Case-II
0/90/0/90/0	10.97430	5.7511
0/90/90/0/0	10.1037	5.2940
0/0/90/90/0	10.1037	5.2940
90/90/0/0/0	5.0320	2.6375

Table 6. reveals the effect of stacking sequence on NCBL of simply supported laminated composite plate with a/h=100, when subjected to biaxial in-plane loading in case-I and case-II. It is clear from the table that plate with symmetric stacking sequence shows maximum NCBL hence have maximum buckling strength. Plate with stacking sequence (0/90/0/90/0) shows 54.14% more value of NCBL than that of plate with stacking sequence (90/90/0/0/0).

4. Conclusion

Behaviour of Laminated composite plate when subjected to biaxial partial in-plane compressive loading for two different loading positions are examined here in the present work. The effect of span to thickness ratio, fiber orientation, stacking sequence and aspect ratio on NCBL are explored. Some important conclusions made from the present work can be given as,

- With increasing dimension of the plate NCBL decreases and plate gets easy to buckle when subjected to biaxial in-plane loading.

- Plate when subjected to biaxial in-plane loading at centre of edge (case-I) is more stable than the plate when subjected to loading at corners of edges.
- With the increment of thickness of the plate gets difficult to buckle.
- Plate with 450 fiber orientation are more stable when subjected to biaxial partial in-plane loading.
- Plate having symmetric stacking sequence is most stable when subjected to biaxial partial in-plane loading.

References

[1] Khayal O and Hussan T 2018 Biaxial buckling of laminated composite plates *International Journal of Bridge Engineering (IJBE),* 19-44.
[2] Khayal O M E S *Biaxial buckling of thin laminated composite plates* Lap Lambert Academic Publishing, ISBN 978-613-8-23615-3.
[3] Libove C 1983 Buckling pattern of biaxially compressed simply supported orthotropic rectangular plates *Journal of composite material*, 17(**1**), 45-48.
[4] Shukla K K, Nath Y, Kreuzer E and Kumar K V S 2005 Buckling of laminated composite rectangular plates *Journal of Aerospace Engineering,* **18**, 215-223.
[5] Tung K and Surdenas J 1987 Buckling of rectangular orthotropic plates under biaxial loading *Journal of Composite Materials,* **21**, 124-128.
[6] Wang X, Wang Y 2015 Buckling analysis of thin rectangular plates under uniaxial or biaxial compressive point loads by the differential quadrature method *International Journal of Mechanical Sciences*, **10**, 07-021.
[7] Spencer H H, Surjanhata H 1986 The simplified buckling criterion applied to plates with partial edge loading *Applied Scientific Research*, **43**, 79-90.
[8] Srivastava A K L, Datta P K, Sheikh A H 2003 Buckling and vibration of stiffened plates subjected to partial edge loading *International Journal of Mechanical Sciences,* **45**, 73-93.
[9] Baker G, Pavlovic M N 1982 Elastic stability of simply supported rectangular plates under locally distributed edge forces *Journal of Applied Mechanics,* **49**/177.

AUTHOR INDEX

Institute of Physics
Dirac House, Temple Back
Bristol BS1 6BE UK

ISSN: 1757-8981
ISBN 978-1-7138-3316-1

CURRAN ASSOCIATES INC.
proceedings.com